Flexible Electronics

Flexible Electronics: Fabrication and Ubiquitous Integration

Special Issue Editor

Ramses V. Martinez

MDPI • Basel • Beijing • Wuhan • Barcelona • Belgrade

MDPI

Special Issue Editor
Ramses V. Martinez
Purdue University
USA

Editorial Office
MDPI
St. Alban-Anlage 66
4052 Basel, Switzerland

This is a reprint of articles from the Special Issue published online in the open access journal *Micromachines* (ISSN 2072-666X) in 2018 (available at: https://www.mdpi.com/journal/micromachines/special_issues/Flexible_Electronics_Fabrication_Ubiquitous_Integration)

For citation purposes, cite each article independently as indicated on the article page online and as indicated below:

LastName, A.A.; LastName, B.B.; LastName, C.C. Article Title. *Journal Name* **Year**, *Article Number*, Page Range.

ISBN 978-3-03897-828-2 (Pbk)
ISBN 978-3-03897-829-9 (PDF)

Contents

About the Special Issue Editor vii

Ramses V. Martinez
Editorial for Special Issue on Flexible Electronics: Fabrication and Ubiquitous Integration
Reprinted from: *Micromachines* **2018**, *9*, 605, doi:10.3390/mi9110605 1

Kangmin Leng, Chuanfei Guo, Kang Wu and Zhigang Wu
Tunnel Encapsulation Technology for Durability Improvement in Stretchable Electronics
Fabrication
Reprinted from: *Micromachines* **2018**, *9*, 519, doi:10.3390/mi9100519 4

Behnam Sadri, Debkalpa Goswami and Ramses V. Martinez
Rapid Fabrication of Epidermal Paper-Based Electronic Devices Using Razor Printing
Reprinted from: *Micromachines* **2018**, *9*, 420, doi:10.3390/mi9090420 16

Lin Xiao, Chen Zhu, Wennan Xiong, YongAn Huang and Zhouping Yin
The Conformal Design of an Island-Bridge Structure on a Non-Developable Surface for
Stretchable Electronics
Reprinted from: *Micromachines* **2018**, *9*, 392, doi:10.3390/mi9080392 30

Bart Plovie, Frederick Bossuyt and Jan Vanfleteren
Stretchability—The Metric for Stretchable Electrical Interconnects
Reprinted from: *Micromachines* **2018**, *9*, 382, doi:10.3390/mi9080382 46

**Jie Tang, Yi-Ran Liu, Li-Jiang Zhang, Xing-Chang Fu, Xiao-Mei Xue, Guang Qian, Ning Zhao
and Tong Zhang**
Flexible Thermo-Optic Variable Attenuator based on Long-Range Surface Plasmon-Polariton
Waveguides
Reprinted from: *Micromachines* **2018**, *9*, 369, doi:10.3390/mi9080369 65

**Tian-Yu Wang, Zhen-Yu He, Lin Chen, Hao Zhu, Qing-Qing Sun, Shi-Jin Ding, Peng Zhou
and David Wei Zhang**
An Organic Flexible Artificial Bio-Synapses with Long-Term Plasticity for Neuromorphic
Computing
Reprinted from: *Micromachines* **2018**, *9*, 239, doi:10.3390/mi9050239 75

**Andrew Stier, Eshan Halekote, Andrew Mark, Shutao Qiao, Shixuan Yang, Kenneth Diller
and Nanshu Lu**
Stretchable Tattoo-Like Heater with On-Site Temperature Feedback Control
Reprinted from: *Micromachines* **2018**, *9*, 170, doi:10.3390/mi9040170 83

Zhaozheng Yu and Huanyu Cheng
Tunable Adhesion for Bio-Integrated Devices
Reprinted from: *Micromachines* **2018**, *9*, 529, doi:10.3390/mi9100529 97

Kyowon Kang, Younguk Cho and Ki Jun Yu
Novel Nano-Materials and Nano-Fabrication Techniques for Flexible Electronic Systems
Reprinted from: *Micromachines* **2018**, *9*, 263, doi:10.3390/mi9060263 116

Yi Ren and Jing Liu
Liquid-Metal Enabled Droplet Circuits
Reprinted from: *Micromachines* **2018**, *9*, 218, doi:10.3390/mi9050218 139

About the Special Issue Editor

Ramses V. Martinez is an assistant professor at the School of Industrial Engineering and the Weldon School of Biomedical Engineering at Purdue University since January 2015. His research group, the FlexiLab, is primarily involved in several projects in the areas of nanomanufacturing, biosensing, and soft robotics. Dr. Martinez is author of more than 40 scientific publications and 15 patents. He has been awarded with the Fulbright Fellowship and the Marie Curie IOF grant for his postdoctoral work in the lab of Prof. George Whitesides at Harvard University while he also served as a Teaching Fellow at the Harvard University Science Center. Dr. Martinez has also been an active mentor of undergrad and graduate students and is now teaching courses on Robotics, Flexible Electronics, Material Science, and Engineering Design at Purdue University. His current research efforts focus in the development of new nanomanufacturing methods for flexible wearable devices and soft robots.

micromachines

MDPI

Editorial

Editorial for Special Issue on Flexible Electronics: Fabrication and Ubiquitous Integration

Ramses V. Martinez [1,2]

[1] School of Industrial Engineering, Purdue University, 315 N. Grant Street, West Lafayette, IN 47907, USA;
rmartinez@purdue.edu
[2] Weldon School of Biomedical Engineering, Purdue University, 206 S. Martin Jischke Drive, West Lafayette,
IN 47907, USA

Received: 26 October 2018; Accepted: 8 November 2018; Published: 19 November 2018

Based on the premise "anything thin is flexible", the field of flexible electronics has been fueled from the ever-evolving advances in thin-film materials and devices. These advances have been complemented by new integration processes that enable the fabrication of bendable, conformable, and stretchable electronic devices over large areas using scalable manufacturing processes. As a result, flexible electronics has underpinned much of the technological innovation in the fields of sensors, solar energy, and displays over the last decades. This Special Issue focuses on the numerous challenges that researchers and engineers must overcome to bring flexible electronic solutions to healthcare, environmental monitoring, and the human–machine interface. The scientific hurdles to overcome affect the design, fabrication, and encapsulation of the flexible electronic devices, making new approaches to improve these fabrication steps to have an immediate impact in the reliable functioning of these devices upon a large range of strains and bending angles. This Special Issue, therefore, brings us one step closer to the expansion of flexible electronic and optical devices for their ubiquitous integration, the development of new form factors, and the opening up of new markets.

There are 10 papers published in this Special Issue, covering new strategies for a paradigm shift in the design [1–3], fabrication [4–7], and encapsulation [8–10] of next-generation flexible systems. Xiao et al. [1] proposed an "island-bridge" strategy to design high-performance stretchable electronics composed of inorganic rigid components so that that can they can be conformally transferred to non-developable surfaces. The design of stretchable electronic devices requires a metric to evaluate their performance. This metric is provided by Plovie et al. [2] to evaluate the performance of stretchable interconnects.

Recent advancements in nanoscale fabrication methods allow the construction of active materials that can be combined with ultrathin soft substrates to form flexible electronics with high performances and reliability. Kang et al. [6] reviewed the most commonly used fabrication methods—involving novel nanomaterials—to make flexible electronics, using application examples of fundamental device components for electronics and applications in healthcare. An alternative, liquid-metal-based soft electronics circuit, termed "droplet circuit" is presented by Ren and Liu [7]. These intrinsically soft circuits can easily match the mechanical impedance of biological tissue and brings significant opportunities for innovation in modern bioelectronics and electrical engineering. A "tunnel encapsulation" strategy is proposed by Leng et al. [8] in order to avoid the typical lack of durability due to stress concentration of flexible interconnects entirely embedded in elastic polymer silicones, such as polydimethylsiloxane (PDMS).

On the application side, these papers have focused on the implementation of flexible systems in healthcare [4,10], photonics [3], and the human–machine interface [9]. Traditional manufacturing approaches and materials used to fabricate flexible epidermal electronics for physiological monitoring, transdermal stimulation, and therapeutics have proven to be complex and expensive, impeding the fabrication of flexible electronic systems that can be used as single-use medical devices. Sadri et al. [4] report the simple, inexpensive, and scalable fabrication of epidermal paper-based electronic devices

(EPEDs) using a bench-top razor printer. These EPEDs are mechanically stable upon stretching and can be used as electrophysiological sensors to record electrocardiograms, electromyograms, and electrooculograms, even under water. Following the trend of fabricating disposable flexible electronic devices for healthcare applications, Stier et al. [10] developed an ultra-soft tattoo-like heater that has autonomous proportional-integral-derivative (PID) temperature control. This epidermal device is capable of maintaining a target temperature typical of medical uses over extended durations of time and to accurately adjust to a new set point in process. The rapid expansion of bio-integrated devices requires the development of new adhesives that will ensure the stability of these systems when implemented over soft biological tissues. Yu and Cheng [5], inspired by the remarkable adhesion properties found in several animal species, review recent developments in the field of tunable adhesives, focusing their applications toward bio-integrated devices and tissue adhesives, where strong adhesion is desirable to efficiently transfer vital signals, whereas weak adhesion is needed to facilitate the removal of those systems.

Tang et al. [3] developed a flexible thermo-optic variable attenuator based on long-range surface plasmon-polariton (LRSPP) waveguide for microwave photonic applications. This flexible plasmonic variable attenuator constitutes a step forward towards the fabrication of high-density photonic integrated circuits and a new solution for data transmission and amplitude control in microwave photonic systems. To improve human–machine interfaces through the construction of neuromorphic computing systems capable of mimicking the bio-synaptic functions, Wang et al. [9] developed a flexible artificial synaptic device with an organic functional layer. This flexible device exhibits retention times of the excitatory and inhibitory post-synaptic currents longer than 60 s.

I would like to take this opportunity to thank all the authors for submitting their papers to this Special Issue. I also want to thank all the reviewers for dedicating their time and helping to improve the quality of the submitted papers.

Conflicts of Interest: The author declares no conflict of interest.

References

1. Xiao, L.; Zhu, C.; Xiong, W.; Huang, Y.; Yin, Z. The Conformal Design of an Island-Bridge Structure on a Non-Developable Surface for Stretchable Electronics. *Micromachines* **2018**, *9*, 392. [CrossRef] [PubMed]
2. Plovie, B.; Bossuyt, F.; Vanfleteren, J. Stretchability—The Metric for Stretchable Electrical Interconnects. *Micromachines* **2018**, *9*, 382. [CrossRef] [PubMed]
3. Tang, J.; Liu, Y.-R.; Zhang, L.-J.; Fu, X.-C.; Xue, X.-M.; Qian, G.; Zhao, N.; Zhang, T. Flexible Thermo-Optic Variable Attenuator based on Long-Range Surface Plasmon-Polariton Waveguides. *Micromachines* **2018**, *9*, 369. [CrossRef] [PubMed]
4. Sadri, B.; Goswami, D.; Martinez, R.V. Rapid Fabrication of Epidermal Paper-Based Electronic Devices Using Razor Printing. *Micromachines* **2018**, *9*, 420. [CrossRef] [PubMed]
5. Yu, Z.; Cheng, H. Tunable Adhesion for Bio-Integrated Devices. *Micromachines* **2018**, *9*, 529. [CrossRef] [PubMed]
6. Kang, K.; Cho, Y.; Yu, K.J. Novel Nano-Materials and Nano-Fabrication Techniques for Flexible Electronic Systems. *Micromachines* **2018**, *9*, 263. [CrossRef] [PubMed]
7. Ren, Y.; Liu, J. Liquid-Metal Enabled Droplet Circuits. *Micromachines* **2018**, *9*, 218. [CrossRef] [PubMed]
8. Leng, K.; Guo, C.; Wu, K.; Wu, Z. Tunnel Encapsulation Technology for Durability Improvement in Stretchable Electronics Fabrication. *Micromachines* **2018**, *9*, 519. [CrossRef] [PubMed]

9. Wang, T.-Y.; He, Z.-Y.; Chen, L.; Zhu, H.; Sun, Q.-Q.; Ding, S.-J.; Zhou, P.; Zhang, D.W. An Organic Flexible Artificial Bio-Synapses with Long-Term Plasticity for Neuromorphic Computing. *Micromachines* **2018**, *9*, 239. [CrossRef] [PubMed]
10. Stier, A.; Halekote, E.; Mark, A.; Qiao, S.; Yang, S.; Diller, K.; Lu, N. Stretchable Tattoo-Like Heater with On-Site Temperature Feedback Control. *Micromachines* **2018**, *9*, 170. [CrossRef] [PubMed]

micromachines

MDPI

Article

Tunnel Encapsulation Technology for Durability Improvement in Stretchable Electronics Fabrication

Kangmin Leng [1], Chuanfei Guo [2], Kang Wu [1],* and Zhigang Wu [1],*

[1]   State Key Laboratory of Digital Manufacturing Equipment and Technology, Huazhong University of Science and Technology, Wuhan 430074, China; lengkangmin@hust.edu.cn

[2]   Department of Materials Science and Engineering, Southern University of Science and Technology, Shenzhen 518055, China; guocf@sustc.edu.cn

*   Correspondence: wuk16@hust.edu.cn (K.W.); zgwu@hust.edu.cn (Z.W.);
    Tel.: +86-150-7239-1101 (K.W.); +86-027-8754-4054 (Z.W.)

Received: 18 September 2018; Accepted: 12 October 2018; Published: 14 October 2018

Abstract: Great diversity of process technologies and materials have been developed around stretchable electronics. A subset of them, which are made up of zigzag metal foil and soft silicon polymers, show advantages of being easy to manufacture and low cost. However, most of the circuits lack durability due to stress concentration of interconnects entirely embedded in elastic polymer silicone such as polydimethylsiloxane (PDMS). In our demonstration, tunnel encapsulation technology was introduced to relieve stress of these conductors when they were stretched to deform in and out of plane. It was realized by dissolving the medium of Polyvinyl Alcohol (PVA), previous cured together with circuits in polymer, to form the micro-tunnel which not only guarantee the stretchability of interconnect, but also help to improve the durability. With the protection of tunnel, the serpentine could stably maintain the designed shape and electrical performance after 50% strain cycling over 20,000 times. Finally, different materials for encapsulation were employed to provide promising options for applications in portable biomedical devices which demand duplicate distortion.

Keywords: stretchable electronics; tunnel encapsulation; Polyvinyl Alcohol; durability

1. Introduction

With a significant development of fabrication processes and patterning technologies, stretchable electronic devices, based on inorganic materials, have achieved high performance in bioinspired and biointegrated systems in the past few years [1–3]. To achieve conformal contact with human skin and maintain stable electrical performance, these devices need to endure strain and stress from many repetitive deformation movements in or out of plane. Special designed structures such as zigzag and serpentine interconnects [4–6] provide the allowance for deformation. This brings the fabrication feasibility for stretchable electronics based on thin inorganic materials [7–9]. Moreover, the prestrain strategy [10] has been reported extensively. It significantly increased the stretchability of interconnects of different shapes for biomedical and healthcare applications has been reported in recent years [11]. Among those studies, very thin metal wires were patterned on the soft substrates through evaporation or transfer printing [12,13]. The finished wires and substrates were encapsulated by pouring a thin soft polymer directly to form a protection layer [5,14,15]. During these processes, encapsulation technologies are closely related to the lifetime and durability when these devices are up against mechanical damage in practical application. The differences of modulus and morphology between metal wires and polymer matrix bring problems when deformation happens. For example, the out-of-plane wavy and 3D buckling deformation for metal wires are constrained by those closely embedded polymers as there are no spaces for these movements [6,14]. They come out of the reducing of desirable stretchability and durability of the systems. Recently, an option involving microfluidic suspensions in thin elastomeric enclosures at system level was proposed [16]. However, this technology was faced with fluid leakage and shape distortion under large deformation. However, research focusing on the stretchability and durability of interconnects with encapsulation needs further supplementary research and perfection [17–19].

Here, we present a new encapsulation technology for the life time improvement of stretchable interconnects. A three-dimensional tunnel space was made to enable slipping and buckling for deformed interconnects. We describe the newly invented encapsulation technology of embedding the serpentine interconnect in three-dimensional tunnel, called tunnel encapsulation. In this study, the micro-tunnel space was realized by dissolving PVA, previous cured together with interconnects in polymer. Nevertheless, patterning the conductive metal films on PVA is a challenge. Previous studies reported some techniques including lithography and ion etching thin films onto PVA [20–22]. These technologies provide patterned films with high accuracy but there are problems such as complicated processes and high cost associated with cleanroom facilities. In addition, greater demands were being placed on the processing equipment, the operator and operating environment. In our technology, metal film and PVA film were directly patterned by laser processing and then bonding with each other for the stickiness of wetting PVA. Experimental studies and finite element analyses (FEA) revealed the important features of our technology and their dependence on key design parameters to predict the durability of the electronic components for a periodic deformation. Different interconnect patterns and polymer matrix materials were applied to evaluate the universality of this new encapsulation and it was proved to be valid. Remarkably, even when stretched with 50% strain, the electrode was still able to recover its original shape and maintain good electrical conductivity. An appropriate dimension and lubricating fluid for the tunnel which enclose the interconnects are critical. In particular, fixing-pillar was proposed to strengthen local stiffness in the arc regions of the serpentine, which could strongly enhance the lifetime of the interconnects. Overall, we provided experimental and simulation results that can fabricate a high-performance electronic device in encapsulation with excellent stretchability and durability. The fundamental difference between the direct encapsulation and tunnel encapsulation is the introduction of the three-dimensional tunnel. For direct encapsulation, interconnect embedded in polymer will encounter severe stress concentration when stretched. The severe stress concentration results in the limited stretchability and lifetime of the interconnect. For tunnel encapsulation, stress concentration is alleviated when stretched at the same degree of elongation for the introduction of tunnel space. Meanwhile, the interconnect would slide and buckle in the tunnel to release stress. With resizing the tunnel, the maximum stretchability and fatigue performance of interconnect can be modified. Therefore, the tunnel encapsulation helps to improve the maximum stretchability and lifetime of the system.

2. Materials and Technologies

2.1. Preparation of PVA Membrane

A 9:1 mixture of deionized water and granular Polyvinyl Alcohol (PVA) (87.0~89.0% hydrolyzed, Mw 130,000 g/mol, Shanghai Macklin Biochemical Co., Ltd., Shanghai, China) was heated in a glass beaker at 75 °C for 1 h with stirring, Then, the mixture was kept at room temperature (25 °C) with vigorous stirring for about 30 min to obtain 10 wt.% PVA aqueous solution. We fabricated pristine PVA membranes by casting the 10 wt.% PVA aqueous solution onto a culture dish and drying the solutions at 75 °C for 2 h.

2.2. Preparation of PDMS

Poly-dimethyl siloxane (PDMS) (Sylgard 186, Dow Corning (Midland, MI, USA)) was chosen as the elastomeric substrate to carry the patterned metal on top. The two-part liquid components (PDMS base and curing agent) were mixed with weight ratio of 10:1 in a plastic cup successively and mixed together manually for 5 min with a glass rod. Prior to putting into use, vacuumization (10 min) and refrigeration (about −18 °C for 1 h) were used to eliminate the air bubbles.

2.3. Fabrication of Stretchable Electronics in Tunnel Encapsulation

A schematic of the tunnel encapsulation process is shown in Figure 1. Firstly, PDMS was poured on top of a ceramic carrier at 90 °C for 10 min to form the donor substrate. In our demonstration, a commercial electrolytic copper foil (EQ-bccf-9u, thinness: 9 μm, Shenzhen Kejing Star Tech, Shenzhen, China) was adhered to the donor substrate. Similarly, two solidified PVA membranes (the width was 1000 μm and the thickness was 50 μm; we dyed the PVA solution with black ink for visibility) were adhered to another two donor substrates. Afterwards, a UV laser marker (HGL-LSU3/5EI, Wuhan Huagong Laser Engineering Co., Ltd., Wuhan, China) was used to pattern the copper Foil and PVA membrane, respectively [23]. The required pattern and unnecessary parts were cut off by UV laser. The patterns of copper and PVA are shown in Figure S1. After tearing off the obsolete PVA membrane and copper foil (Figure 1a), a patterned PVA film was transferred from a donor to a receiver substrate (made at 90 °C for 4 min) for the different stickiness of substrate. Then, the required copper circuit was wetted with deionized water (place the copper over a bottle filled with 100 °C deionized water about 15 s) and put into contact with the copper and PVA without external pressure for 5 s. Afterwards, the copper circuit was sticking to the patterned PVA due to stickiness increasing of wetting PVA (Figure 1b). In other words, the copper was transferred from the donor to the patterned PVA which was on the receiver substrate. Same as above procedure, another donor substrate covered with patterned PVA was wetted with deionized water, and the patterned PVA was transferred from the donor to the top of the double layer. After stacking up, the water would dissolve the interface of patterned PVA. The copper was embedded in PVA when constructing the PVA/copper/PVA sandwich structure (Figure 1c). Then, the PDMS was poured to directly encapsulate the sandwich structure and cured for 2 h at 75 °C (Figure 1d). The PDMS/sandwich structure/PDMS laminate was cut into strips, each strip has one patterned meander metal interconnect encapsulated in the stretchable substrate. The edges of strips were cut to expose the entrance and exit. At the final step, these strips were placed in deionized water for 24 h at 75 °C to dissolve the PVA which enclosed those copper circuits (Figure 1e). The residual PVA was washed away by an injection needle filled with deionized water. From a practical point of view, the entrance and exit of these strips should be sealed to prevent the corrosion of the circuits. The critical point of the fabrication is the intimate bonding of PVA and copper, and it can be assured by vacuuming when pouring the upper PDMS before curing. Nevertheless, it may affect the shaping of tunnel.

2.4. Stretching and Electrical Test

All the strips were clamped by two pieces of thin PDMS on both sides, and a high-frequency fatigue testing machine (E1000, Instron, Boston, MA, USA) stretched the sample to a specified displacement specified through program for cyclic stretching at 2 Hz. The ends of the interconnects were welded to connect the test electrode; a digital multimeter (34,461A, KeySight Technologies, Santa Rosa, CA, USA) was used for electrical test.

2.5. Finite Element Analysis

ABAQUS commercial software (ABAQUS6.14, ABAQUS Inc., Palo Alto, CA, USA) was used to study the mechanics response of tunnel encapsulation and direct encapsulation. PDMS was modeled by the hexahedron element (C3D8R), while the interconnect was modeled by the shell element (S4R). 22,749 elements for silicone and 1419 elements for interconnects were used to conduct FEA modeling after grid independence testing. Displacement boundary conditions were applied to both edges of the system to apply different levels of stretching and refined meshes were adopted to ensure the accuracy. ABAQUS/Explicit was applied to analyze the deformation and normal stress distribution of interconnects with tunnel encapsulation and direct encapsulation.

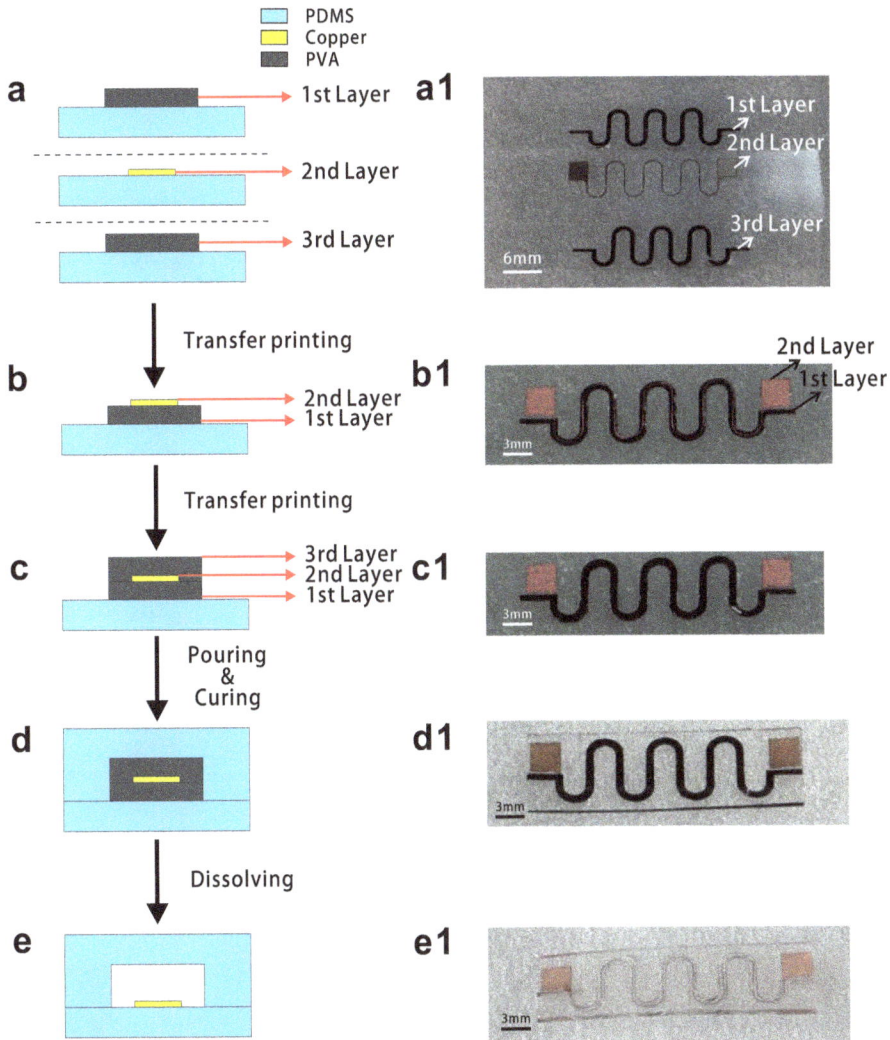

Figure 1. Schematic diagram of the tunnel encapsulation fabrication process. (**a**) Patterned copper foil and PVA membrane, where the first layer and the third layer have the same dimension. (**a1**) Optical image of patterned copper film and PVA. (**b**) Metal circuit was transferred to the patterned PVA. (**b1**) Optical image of the double layer on the receiver substrate. (**c**) Another patterned PVA was transferred to the top of the double layer structure. (**c1**) Optical image of the sandwich structure. (**d**) PDMS was poured to direct encapsulate the sandwich structure and cured at 75 °C for 2 h. (**d1**) Optical image of the sandwich structure encapsulated with PDMS. (**e**) Both ends of the laminate strip were cut to expose the entrance and exit for dissolving the PVA around the Copper circuit. (**e1**) Optical image of the sample after dissolving the PVA.

3. Results and Discussion

3.1. The Performance during Stretching

Figure 2a shows three different patterns of serpentine interconnects which consist of two periodic unit cells of two straight lines and two arc of circles, and the straight lines are tangent to the arc. R is the outer radius, S is the space of two unit cells, L is the center distance of two arc along the longitudinal axis, and W is the width of the copper trace. The center distance of these patterns was fixed to set a limit for the dimensions of these circuits. Therefore, R is the critical parameter of these pattern, which controls the shape of the pattern. We obtained horseshoe design (Pattern-A, R = 1.95 mm, S = 6 mm, L = 4 mm, W = 100 μm), semicircle design (Pattern-B, R = 1.55 mm, S = 6 mm, L = 4 mm, W = 100 μm) and the snakelike design (Pattern-C, R = 1.15 mm, S = 6 mm, L = 4 mm, W = 100 μm). Let Ht denote the height of the tunnel made of PVA; Wt the width of the tunnel; Ws the width of the strip; and Ts the thickness of the strip. Through the thickness, the strip has a layered structure (Figure 2b) from bottom (substrate/serpentine/encapsulation layer) to top. A sample (Ht = 100 μm, Wt = 1000 μm, Ws = 1 cm, Ts = 500 μm) with serpentine designed Pattern-B is presented sequentially to illustrate the characteristic of tunnel encapsulation. Figure 2c shows the original shape of the interconnect with tunnel encapsulation before stretched and the tunnel was filled with green fluid to identify the deformation of tunnel. Figure 2d exhibits the serpentine interconnect at 30% applied strain obtained from dynamic mechanical analysis. Figure 2e exhibits the serpentine interconnect at 50% applied strain. Figure 2g illustrates the interconnects slip and buckle in the tunnel when stretched. It is shown that buckling occurred in the arc region highlight with the yellow frame and it is identical with the FEA result (magnification in Figure 2f). The corresponding FEA predictions on buckling deformation of serpentine interconnects illustrate that the tunnel provides enormous space for the interconnects to relieve stress with slipping and buckling.

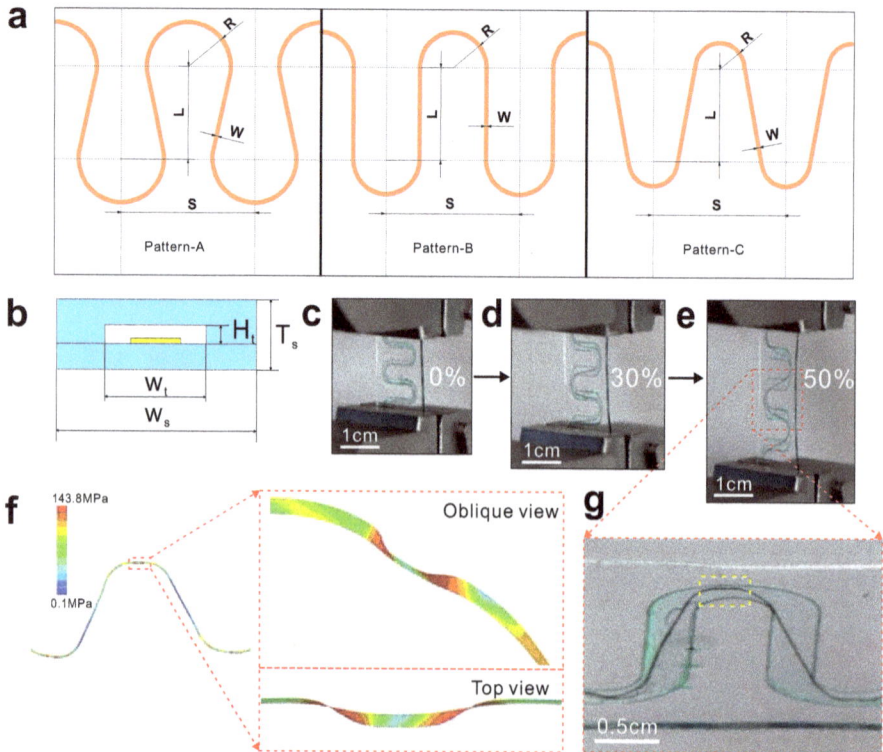

Figure 2. (**a**) Three different patterns of serpentine interconnects. (**b**) Cross-sectional illustration of representative layers in a hollow structure with embedded stretchable interconnects which collapse in the tunnel. (**c–e**) Optical images of the sample before stretched, morphology at 30% and 50% applied strain, respectively. (**f**) FEA results of morphology at 50% strain. (**g**) Magnification of slipping and buckling region.

We performed elongation test with both encapsulation technologies to investigate the effect of encapsulation technology on maximum stretchability. Three serpentines of different patterns were encapsulated in PDMS with both technologies to confirm the effect of patterns on maximum stretchability. In addition, to evaluate the effect of thickness of encapsulation layer, we made samples with three different thicknesses (300 μm, 500 μm, and 700 μm). The results of the experiment are shown in Figure 3a, where the direct encapsulation and tunnel encapsulation are abbreviated to DE and TE, respectively. For direct encapsulation without further processing, the maximum elongation is limited by localized fracture of the copper embedded in silicone, and the interconnect encounters severe stress concentration originating from constraints of surrounding PDMS (Figures S2 and S3). The FEA result in Figure 3e illustrates the stress of interconnect with direct encapsulation increase dramatically (stress at 80–110% strain exceed the maximum stress, hence values are not shown in the graph). For tunnel encapsulation, the serpentine interconnect was freestanding in the tunnel except the two ends. Upon stretching of the whole system, the interconnect is dragged from the two ends, geometrical opening in arc segment increases to accommodate the elongation, the straight part and the arc segment undergoes both stretching and bending. As the applied strain exceeds a critical value, lateral buckling occurs to reduce the strain energy, which helps to alleviate stress concentration of interconnects. Figure 3e shows that the stress of interconnects with tunnel encapsulation is considerably lower than interconnect with direct encapsulation. Besides, the platform period of stress with tunnel encapsulation in Figure 3e proved that the outer stretching is accommodated by the sliding and buckling behavior of interconnect in tunnel (Figures S4–S6). It is shown that Pattern-C owns the best stretchability in both encapsulation technologies. A larger curvature in the arc region become closer to a straight line. That means Pattern-C serpentine has a larger redundancy to deform with both technologies as the curvature of Pattern-C is the smallest. The stretchability of the system decreases as the thickness of the strip increases when encapsulated directly (Figure 3a) since the interconnects suffer severer mechanical constraints at elongation. On the contrary, the trend of the curve of tunnel encapsulation is opposite to the direct encapsulation. In tunnel encapsulation, PDMS around the serpentine broke first before the copper rupture as the tunnel would decrease the strength of the strip. As a result, with the same dimension of tunnel, the degradation of a thinner strip will become worse.

Figure 3b illustrates the reversibility of serpentine (Pattern-B, Ht = 100 μm, Wt = 1000 μm, Ws = 1 cm, Ts = 500 μm) in tunnel encapsulation. The serpentine can recover its original shape roughly under 50% strain after 20,000 cycles at 2 Hz. As the interconnects hang in the air deformed, typically by out-of-plane buckling and in-plane bending, they will lose the original shape when the stress is relieved. With tunnel encapsulation, the surrounding silicone matrix will mechanically constrain deformation of the interconnects in the case of the serpentine coming loose in the tunnel space. In the experiment, as long as the applied strain was under maximum strain for all kinds of patterns and thicknesses, the system could ensure reversibility in its mechanics.

3.2. The Effect of Stress Concentration on Durability

We studied the influence of our technology on the durability with the same pattern design (Pattern-B). We performed cyclic loading experiments with a focus on the lifetime of the serpentine, defined as the number of stretching cycles to failure in the interconnect. Figure 3c shows the number of stretching cycles to failure in the interconnect versus applied strain for different encapsulation technologies and different system thicknesses (300 μm, 500 μm, and 700 μm). For the direct encapsulation, the differences on lifetime with corresponding elongation are caused by the fact that the encapsulation would further constrain the interconnects from deforming out of plane if elongation is increased. Meanwhile, stress concentration occurs in the arc region of the serpentine and the phenomenon becomes rather obvious with increasing elongation. The FEA result (Figure 3e,f) also verifies the hypothesis. Remarkably, the lifetime of serpentine improved tremendously by the tunnel encapsulation. Likewise, the differences on lifetime with corresponding elongation are caused by further constraints and severer friction contact. Note that the outstanding improvement on lifetime with tunnel encapsulation stem from the introduction of tunnel space. It offers the activity space for the interconnects to slide and buckle in the space to relieve stress at elongation. In addition, the tunnel isolates the interconnect from surrounding PDMS and helps to decrease the friction of interconnect when stretching. The result in Figure 3c reveals that a thinner system would improve the lifetime of the serpentine as the constraints of a thinner encapsulation layer are weaker. It can be concluded from those results that the lifetime of the serpentine decreases significantly with increasing system elongation. Besides, both the tunnel encapsulation technology and a thinner encapsulation layer help to improve the lifetime of the interconnects.

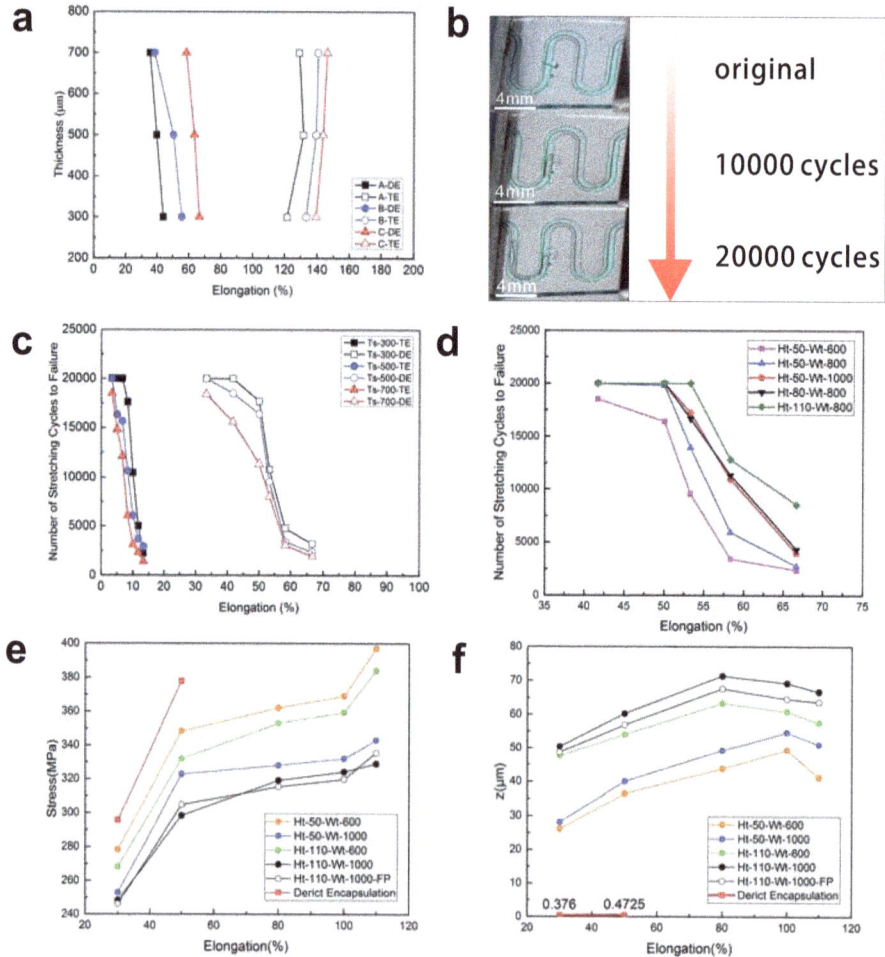

Figure 3. (**a**) Maximum stretchability when the serpentine (different patterns) encapsulated in different technologies with different sample thicknesses Ts from 300 to 700 μm and the width of these samples is 600 μm. (**b**) Reversibility illustration of serpentine in tunnel encapsulation (Ht = 100 μm, Wt = 1000 μm, Ws = 1 cm, Ts = 500 μm). (**c**) Number of stretching cycles to failure when the serpentine encapsulated in different technologies with different sample thicknesses Ts from 300 to 700 μm and the width of these samples is 600 μm. (**d**) Experimental results for different width and height of tunnel in system (Ts = 500 μm). (**e**,**f**) FEA predictions on stress and relative displacement in z direction of serpentine interconnects under stretching.

The available space depends on the width and the height of the tunnel. Thus, a comparison of different width tunnel encapsulation is presented in Figure 3d. It is obvious that the lifetime increase with a wider tunnel. It might be attributed to a broader space enabling a greater degree of slipping and buckling of the interconnects (see green and black line in Figure 3f). In addition, the result depicts the relationship between lifetime of interconnects and the height of tunnel. The lifetime increased as the height increased, which gives rise to a larger space to buckle for the serpentine in the tunnel under stretching, which is a desirable characteristic for cycling stretch. The FEA result (see orange and green line, blue and black line in Figure 3f) shows that the greater heights enable the interconnects to have a larger displacement in z direction, which means the stress relieved when buckling in the direction (see orange and green line, blue and black line in Figure 3e). The displacement in z direction would decrease after a specified value of strain for the severer buckling (Supplementary Materials, Figure S7). Therefore, the durability of the interconnects can be improved by increasing the width and height of the tunnel.

The tunnel was filled with a silicone oligomer (Sylgard 184, without curing agent) using an injection needle to lubricate the contact surface and reduce the friction between interconnects and silicon. The effect of injected fluid was quantitatively studied by experimental tests, as shown in Figure 4a. It is clear that the lifetime of the interconnects was improved as the lubricating fluid reduced the nonspecific adhesion between the interconnects and the tunnel. Besides, it is hydrophobic, expelling moisture from the package, and is optically transparent. It is remarkable that all the experiments were done at once after dissolving. It can be seen that the lifetime of interconnect in water-filled tunnel declines rapidly and the water may corrode the copper interconnect. On the contrary, interconnect in silicone oligomer-filled tunnel sustains excellent fatigue performance without the risk of corrosion. The contact between the interconnects and the wall of the tunnel will definitely influence the lifetime of the serpentine. Apparently, the height of the tunnel plays an important role in optimization as described above. Taking the interconnect itself into account, diminishing the out-of-plane displacement when stretched may contribute much to the lifetime of the system. Here, we made several flexible fixing-pillars in the tunnel to fix the position of the interconnect. The results are shown in Figure 4b. It is important to note that the experiment is based on the lubricating fluid. It was observed that the lifetime increased enormously with introduction of fixing-pillars. We reduced the local stiffness of the arc region of the serpentine so that the deformation in the arc part would be alleviated, and the probability of contacting between interconnects and the wall of the tunnel diminished accordingly. Of course, the pillars are located in the minimum deformation part of the interconnects. For the patterns we used, it was the middle part of the straight segments of the serpentine structure. The FEA result shows that the fixing-pillar help to decrease displacement in z direction while maintaining similar degree of stress (Figure 3e,f). Besides, it just needs to modify the pattern of the PVA a little bit to be an effective technology that is able to improve the durability of interconnects when applying this new technology of encapsulation.

3.3. Electrical Performance

The stretchable serpentine shows excellent cycling stability in Figure 4c. The parameters of the sample are Ts = 500 μm, Wt = 100 μm, and Ht = 100 μm, after 20,000 cycles at 50% applied strain, the interconnects still exhibited an almost constant value. Note that no visible cracks formed in the interconnects when inspected under an optical microscope, and the system can maintain stable electrical performance under repetitively large deformation.

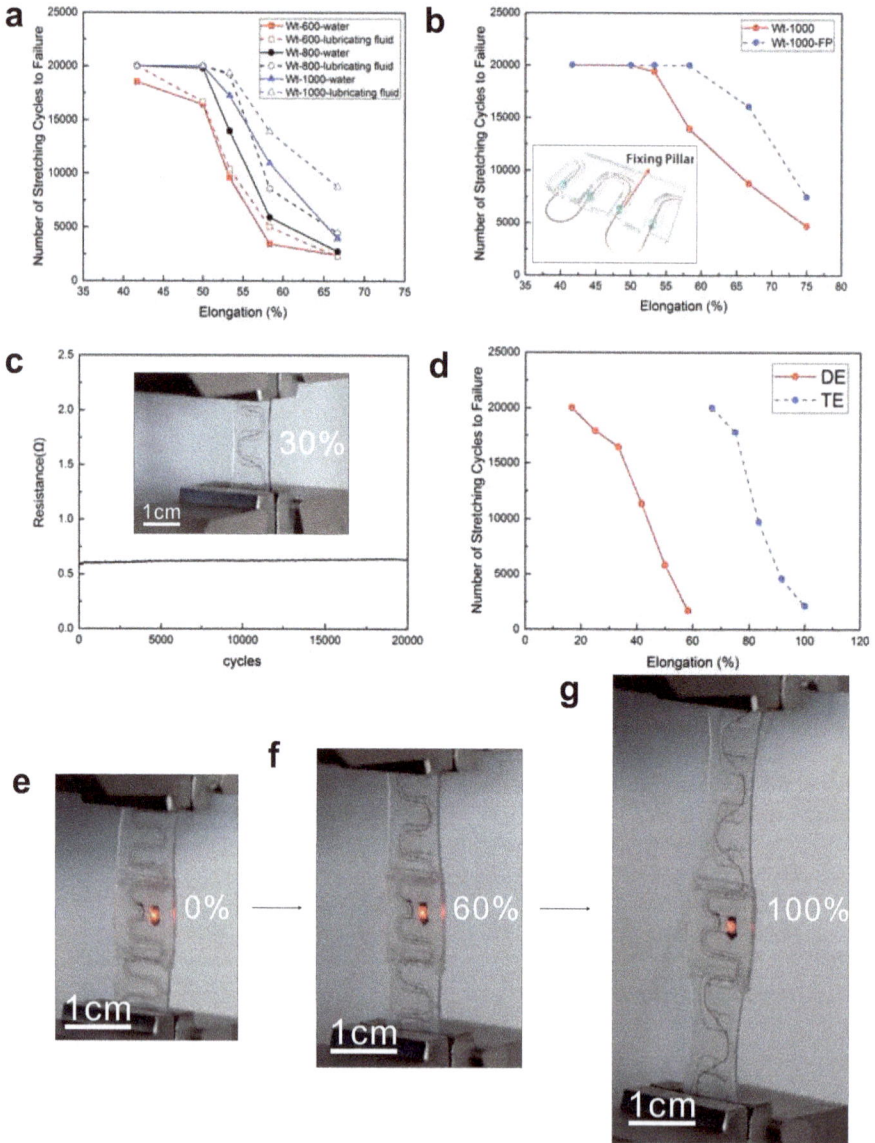

Figure 4. (**a**) Experimental results for effect of lubricating fluid on lifetime of samples with different tunnel widths (sample thickness Ts = 500 μm and height of the tunnel Ht = 100 μm). (**b**) The corresponding improvement in lifetime when adopt fixing-pillars (300 μm in diameter) in the tunnel filled with lubrication fluid (Ht = 100 μm, Wt = 1000 μm, Ws = 1 cm, Ts = 500 μm) and schematic illustration of fixing-pillars (insets). (**c**) Experimental results of resistance change with fixing-pillars in tunnel (Ht = 100 μm, Wt = 1000 μm, Ws = 1 cm, Ts = 500 μm) and optical images of deformation with pillars in tunnel at 30% strain. (**d**) Experimental results of serpentine with direct encapsulation and tunnel encapsulation in Ecoflex. (**e**–**g**) Optical images of device before stretched and morphology at 60% and 100% applied strain, respectively.

Micromachines **2018**, *9*, 519

3.4. The Universality of Tunnel Encapsulation Technology

To study the universal property of this new encapsulation, Ecoflex was selected as encapsulation material to test the durability of the system in Figure 4d. The lifetime of the serpentine encapsulated with Ecoflex improved enormously compared with PDMS, and it is suitable for tunnel encapsulation. In addition, it was observed that the interconnects reached its rupture strain before the Ecoflex broke. The stretchability is up to 200% for the serpentine structure of the metal, but not the ultimate limit for interconnects which are not encapsulated. The interconnects in the tunnel can recover the original shape practically even at the maximum elongation for the constraints of the tunnel.

4. Device Display

A long serpentine interconnect encapsulated in Ecoflex with tunnel encapsulation is presented in Figure 4e–g. An LED was integrated in the device, and fixing-pillars were applied in the tunnel. The parameters of the device are Ts = 500 µm, Wt = 1000 µm, and Ht = 150 µm, and the interconnect with tunnel encapsulation can sustain electrical performance when stretched to reach 100% strain. In this device, a piece of PDMS was attached over the central part to protect the connecting of the LED chip and interconnects. When the system was elongated, the interconnects of both end endure the whole elongation while the middle part would sustain the original shape. Considering integration of microchip, it is worth exploring placing a microchip in the three-dimensional tunnel to further alleviate stress concentration on the chip.

5. Conclusions

In summary, we report a technology of tunnel encapsulation to improve the lifetime of stretchable interconnects A tunnel is formed from dissolution of PVA in deionized water which can help alleviate the stress concentration by provide the space for the sliding and buckling of interconnects. Our tunnel encapsulation confers superior properties compared to direct encapsulation, namely: (1) exceptional stretchability and durability, our approach offers a three-dimensional activity space for the interconnects to buckle, twist and stretch, the stretchability can be up to 200% when the interconnects encapsulated in Ecoflex; (2) excellent stability of electronics conductivity, the interconnects can sustain excellent electrical performance after 20,000 cycles at 50% applied strain; and (3) ease of patterning, the pattern of the tunnel can be readily changed by modifying the pattern of PVA. This new encapsulation has a good application prospect for consumer wearable electronics.

Supplementary Materials: The following are available online at http://www.mdpi.com/2072-666X/9/10/519/s1, Figure S1: (a) Pattern of copper; (b) Pattern of Polyvinyl Alcohol (PVA), Figure S2: Stress distributions and strain contours of the interconnects with direct encapsulation at 30% elongation, Figure S3: Stress distributions and strain contours of the interconnects with direct encapsulation at 50% elongation, Figure S4: Stress distributions and strain contours of the interconnects with tunnel encapsulation at 30% elongation, Figure S5: Stress distributions and strain contours of the interconnects with tunnel encapsulation at 50% elongation, Figure S6: Stress distributions and strain contours of the interconnects with tunnel encapsulation at 100% elongation, Figure S7: Stress distributions and strain contours of the interconnects with tunnel encapsulation at 110% elongation.

Author Contributions: K.W. and K.L. conceived the concept and designed the experiments. K.L. conducted the devices fabrication, testing and FEA modeling. K.L., K.W., and G.C. conducted the data analysis. Z.W. supervised and supported the work. All authors contributed to the manuscript and revision.

Funding: This research received no external funding.

Acknowledgments: This work was supported by the National Key Research and Development Program of China (2017YFB1303103), National Natural Science Foundation of China (No. U1613204) and China Postdoctoral Science Foundation (2018M632833). Wu and Guo thank the support from Chinese central government through its Thousand Youth Talents program.

Conflicts of Interest: The authors declare no conflict of interest.

Micromachines **2018**, *9*, 519

References

1. Sun, Y.; Kumar, V.; Adesida, I.; Rogers, J.A. Buckled and wavy ribbons of GaAs for high-performance electronics on elastomeric substrates. *Adv. Mater.* **2006**, *18*, 2857–2862. [CrossRef]
2. Sun, Y.; Rogers, J.A. Structural forms of single crystal semiconductor nanoribbons for high-performance stretchable electronics. *J. Mater. Chem.* **2007**, *17*, 832–840. [CrossRef]
3. Kim, D.H.; Kim, Y.S.; Wu, J.; Liu, Z.; Song, J.; Kim, H.S.; Huang, Y.Y.; Hwang, K.C.; Rogers, J.A. Ultrathin silicon circuits with strain-isolation layers and mesh layouts for high-performance electronics on fabric, vinyl, leather, and paper. *Adv. Mater.* **2009**, *21*, 3703–3707. [CrossRef]
4. Zhao, Y.; Huang, X. Mechanisms and materials of flexible and stretchable skin sensors. *Micromachines* **2017**, *8*, 69. [CrossRef]
5. Gonzalez, M.; Axisa, F.; Bossuyt, F.; Hsu, Y.Y.; Vandevelde, B.; Vanfleteren, J. Design and performance of metal conductors for stretchable electronic circuits. *Circuit World* **2009**, *35*, 22–29. [CrossRef]
6. Kim, D.H.; Liu, Z.; Kim, Y.S.; Wu, J.; Song, J.; Kim, H.S.; Huang, Y.; Hwang, K.C.; Zhang, Y.; Rogers, J.A. Optimized structural designs for stretchable silicon integrated circuits. *Small* **2009**, *5*, 2841–2847. [CrossRef] [PubMed]
7. Li, Z.; Wang, Y.; Xiao, J. Mechanics of curvilinear electronics and optoelectronics. *Curr. Opin. Solid State Mater. Sci.* **2015**, *19*, 171–189. [CrossRef]
8. Li, Z.; Wang, Y.; Xiao, J. Mechanics of bioinspired imaging systems. *Theor. Appl. Mech. Lett.* **2016**, *6*, 11–20. [CrossRef]
9. Li, Z.; Xiao, J. Mechanics and optics of stretchable elastomeric microlens array for artificial compound eye camera. *J. Appl. Phys.* **2015**, *117*, 014904. [CrossRef]
10. Sun, Y.; Choi, W.M.; Jiang, H.; Huang, Y.Y.; Rogers, J.A. Controlled buckling of semiconductor nanoribbons for stretchable electronics. *Nat. Nanotechnol.* **2006**, *1*, 201–207. [CrossRef] [PubMed]
11. Yeo, W.H.; Kim, Y.S.; Lee, J.; Ameen, A.; Shi, L.; Li, M.; Wang, S.; Ma, R.; Jin, S.H.; Kang, Z.; et al. Multifunctional epidermal electronics printed directly onto the skin. *Adv. Mater.* **2013**, *25*, 2773–2778. [CrossRef] [PubMed]
12. Meitl, M.A.; Zhu, Z.T.; Kumar, V.; Lee, K.J.; Feng, X.; Huang, Y.Y.; Adesida, I.; Nuzzo, R.G.; Rogers, J.A. Transfer printing by kinetic control of adhesion to an elastomeric stamp. *Nat. Mater.* **2006**, *5*, 33–38. [CrossRef]
13. Carlson, A.; Bowen, A.M.; Huang, Y.; Nuzzo, R.G.; Rogers, J.A. Transfer printing techniques for materials assembly and micro/nanodevice fabrication. *Adv. Mater.* **2012**, *24*, 5284–5318. [CrossRef] [PubMed]
14. Hsu, Y.Y.; Gonzalez, M.; Bossuyt, F.; Axisa, F.; Vanfleteren, J.; De Wolf, I. The effects of encapsulation on deformation behavior and failure mechanisms of stretchable interconnects. *Thin Solid Films* **2011**, *519*, 2225–2234. [CrossRef]
15. Jablonski, M.; Lucchini, R.; Bossuyt, F.; Vervust, T.; Vanfleteren, J.; De Vries, J.; Vena, P.; Gonzalez, M. Impact of geometry on stretchable meandered interconnect uniaxial tensile extension fatigue reliability. *Microelectron. Reliab.* **2015**, *55*, 143–154. [CrossRef]
16. Xu, S.; Zhang, Y.; Jia, L.; Mathewson, K.E.; Jang, K.I.; Kim, J.; Fu, H.; Huang, X.; Chava, P.; Wang, R.; et al. Soft microfluidic assemblies of sensors, circuits, and radios for the skin. *Science* **2014**, *344*, 70–74. [CrossRef] [PubMed]
17. Hussain, A.M.; Ghaffar, F.A.; Park, S.I.; Rogers, J.A.; Shamim, A.; Hussain, M.M. Metal/polymer based stretchable antenna for constant frequency far-field communication in wearable electronics. *Adv. Funct. Mater.* **2015**, *25*, 6565–6575. [CrossRef]
18. Pan, T.; Pharr, M.; Ma, Y.; Ning, R.; Yan, Z.; Xu, R.; Feng, X.; Huang, Y.; Rogers, J.A. Experimental and theoretical studies of serpentine interconnects on ultrathin elastomers for stretchable electronics. *Adv. Funct. Mater.* **2017**, *27*, 1702589. [CrossRef]
19. Zhang, Y.; Wang, S.; Li, X.; Fan, J.A.; Xu, S.; Song, Y.M.; Choi, K.J.; Yeo, W.H.; Lee, W.; Nazaar, S.N.; et al. Experimental and theoretical studies of serpentine microstructures bonded to prestrained elastomers for stretchable electronics. *Adv. Funct. Mater.* **2014**, *24*, 2028–2037. [CrossRef]
20. Le Borgne, B.; De Sagazan, O.; Crand, S.; Jacques, E.; Harnois, M. Conformal Electronics Wrapped Around Daily Life Objects Using an Original Method: Water Transfer Printing. *ACS Appl. Mater. Interfaces* **2017**, *9*, 29424–29429. [CrossRef] [PubMed]

Micromachines **2018**, *9*, 519

21. Rogel, R.; Borgne, B.L.; Mohammed-Brahim, T.; Jacques, E.; Harnois, M. Spontaneous Buckling of Multiaxially Flexible and Stretchable Interconnects Using PDMS/Fibrous Composite Substrates. *Adv. Mater. Interfaces* **2017**, *4*, 1600946. [CrossRef]
22. Jin, S.H.; Kang, S.K.; Cho, I.T.; Han, S.Y.; Chung, H.U.; Lee, D.J.; Shin, J.; Baek, G.W.; Kim, T.i.; Lee, J.H.; et al. Water-soluble thin film transistors and circuits based on amorphous indium–gallium–zinc oxide. *ACS Appl. Mater. Interfaces* **2015**, *7*, 8268–8274. [CrossRef] [PubMed]
23. Peng, P.; Wu, K.; Lv, L.; Guo, C.F.; Wu, Z. One-Step Selective Adhesive Transfer Printing for Scalable Fabrication of Stretchable Electronics. *Adv. Mater. Technol.* **2018**, *3*, 1700264. [CrossRef]

micromachines

MDPI

Article

Rapid Fabrication of Epidermal Paper-Based Electronic Devices Using Razor Printing

Behnam Sadri [1] , **Debkalpa Goswami** [1] **and Ramses V. Martinez** [1,2,*]

1   School of Industrial Engineering, Purdue University, 315 N. Grant Street, West Lafayette, IN 47907, USA;
    bsadri@purdue.edu (B.S.); dgoswami@purdue.edu (D.G.)
2   Weldon School of Biomedical Engineering, Purdue University, 206 S. Martin Jischke Drive, West Lafayette,
    IN 47907, USA
*   Correspondence: rmartinez@purdue.edu; Tel.: +1-765-496-0399

Received: 12 May 2018; Accepted: 7 June 2018; Published: 22 August 2018

Abstract: This work describes the use of a benchtop razor printer to fabricate epidermal paper-based electronic devices (EPEDs). This fabrication technique is simple, low-cost, and compatible with scalable manufacturing processes. EPEDs are fabricated using paper substrates rendered omniphobic by their cost-effective silanization with fluoroalkyl trichlorosilanes, making them inexpensive, water-resistant, and mechanically compliant with human skin. The highly conductive inks or thin films attached to one of the sides of the omniphobic paper makes EPEDs compatible with wearable applications involving wireless power transfer. The omniphobic cellulose fibers of the EPED provide a moisture-independent mechanical reinforcement to the conductive layer. EPEDs accurately monitor physiological signals such as ECG (electrocardiogram), EMG (electromyogram), and EOG (electro-oculogram) even in high moisture environments. Additionally, EPEDs can be used for the fast mapping of temperature over the skin and to apply localized thermotherapy. Our results demonstrate the merits of EPEDs as a low-cost platform for personalized medicine applications.

Keywords: epidermal sensors; stretchable electronics; wireless power; hydrophobic paper; wearable stimulators; paper electronics; low-cost manufacture

---

1. Introduction

The ever-growing demand for wearable technologies capable of monitoring key physiological signals have a predicted market growth from \$15b in 2015 to \$150b in 2027 [1]. Wearable healthcare devices fabricated using conventional rigid platforms and silicon-based technologies have been demonstrated to be useful in the continuous collection of clinically relevant personalized information for diseases such as heart failure [2], erythema [3], and diabetes [4,5]. Unfortunately, these rigid or semi-rigid wearable devices can lead to inconsistent measurements due to their limited conformability to human skin during motion and they are often perceived by patients as uncomfortable, thus hindering their adoption.

A new class of thin, flexible, and stretchable electronics, known as epidermal electronic systems [6], has emerged as wearable healthcare tools capable of efficiently monitoring a variety of physiological signals [7] and stimulating different tissues [8]. The stretchability and low mechanical impedance (similar to human skin) make epidermal electronics suitable for continuous health monitoring using wearable devices, due to their intimate contact with the skin and their compliance to its natural moves.

Thin films of ductile metals such as gold or platinum have been extensively used in the fabrication of contact electrodes for epidermal electronics due to their chemical stability and low resistivity [9]. Several electrode designs such as serpentine patterns [10], fractal designs [11], and self-similar buckles [12] have been explored to match the mechanical impedance of human skin.

The manufacture of epidermal electronics with conformable designs often require clean room processes, such as photolithography [13], wet etching [9,14], and physical vapor deposition [7] to create conductive electrodes capable of conforming to the skin. A variety of epidermal electronic systems assembled on flexible substrates have demonstrated excellent measuring performances even during stretching or severe bending, with a resolution comparable to that of advanced CMOS technologies [15]. Unfortunately, the high cost of the materials and the complexity of the fabrication processes (incompatible with large scale manufacturing) required to manufacture epidermal electronics make them unsuitable for personalized medicine applications.

Paper has become a popular substrate for flexible electronics due to its printability, low cost, light weight, and disposability [16]. Conductive inks [17,18], semiconductors [19,20], and insulators [21] can be introduced to papers to tailor the electrical properties of the final device using simple manufacturing processes such as inkjet printing [19,22], spin coating [23], or screen-printing [24] to make devices such as transistors [25], batteries [26], solar cells [27], light-emitting diodes [28], triboelectric generators [29], and antennas [30]. The limited stretchability (below 5%) of paper, however, makes paper-based electronics unsuitable for epidermal applications. Moreover, the performance of electronic devices printed on conventional paper is sensitive to relative humidity and temperature. The development of a simple method to fabricate a variety of paper-based epidermal electronic devices will be desirable to significantly reduce their cost and manufacturing time.

Several approaches have been proposed to change the wetting properties of paper to improve its electrical stability and mechanical integrity in high humidity environments [31]. Infusing hydrophobic materials such as wax or photoresists has enabled the selective modification of the wettability of the paper, enabling the fabrication of low-cost microfluidic devices [32,33]. Recently, our group introduced the selective functionalization of the cellulose fibers of paper using fluoroalkyl trichlorosilanes (R^F) as a fast and simple way to render paper omniphobic, limiting its wettability by aqueous solutions and non-polar solvents [34]. Omniphobic paper reduces the consumption of conductive ink during printing processes, reducing the price of printed electronics, and avoids the degradation of the mechanical properties of the paper due to environmental humidity. The moisture insensitivity and light weight of omniphobic paper has promoted its use as a low-cost substrate for applications in MEMS [35], microfluidics [36], and portable analytical devices [37]. Additionally, omniphobic paper devices can be easily prototyped using a variety of scalable tools such as engravers, laser cutters, and razor printers [38].

Our previous work on paper-based microfluidics demonstrated the low-cost fabrication of analytical systems by employing a thin cutting blade [6]. Here, we demonstrate the use of razor printing to rapidly fabricate epidermal paper-based electronic devices (EPEDs). EPEDs described in this study comprise a layer of a conductive material (metallic thin film or microparticle-based ink) attached to a layer of cellulose paper rendered omniphobic through silanization. Razor printed EPEDs can be rapidly fabricated at a low cost and offer several advantages as follows: (i) They are lightweight, thin, flexible, and even more stretchable than human skin; (ii) They are capable of real-time monitoring of bio-signals such as electrocardiogram (ECG), electro-oculogram (EOG), and electromyogram (EMG) with high precision, independently of environmental moisture or sweating of the wearer; (iii) The wide variety of thin metallic films and conductive inks compatible with razor printing provides an ample range of conductive agents to tailor the functionality of the EPED; (iv) EPEDs fabricated with flexible conductive inks can be used to monitor temperature and provide localized heat therapy, and (v) The omniphobic fibers of the paper reinforce the conductive layer of the EPEDs, preserving the electrical conductivity of the device upon stretching and making these devices compatible with wireless power transfer applications.

2. Materials and Methods

2.1. Choice of Materials

We purchased Whatman#1 paper (GE Healthcare Inc., Philadelphia, PA, USA) and thin paper (70-μm-thick, Elements 300, amazon.com) to serve as substrates for the EPEDs. Thin copper foils (20-μm-thick, Kraftex Products, Gloucestershire, UK) and Ag/AgCl ink (AGCL-675, Applied Ink Solutions, Westborough, MA, USA) were employed as conductive layers. We used a solution of a long-chain fluorinated organosilane (Diisopropyl(3,3,4,4,5,5,6,6,7,7,8,8,9,9,10,10,10-heptadecafluorodecyl)silane, Sigma-Aldrich Corp., St. Louis, MO, USA) to render the paper substrates omniphobic.

2.2. Fabrication of EPEDs by Razor Printing

We functionalized the paper substrates by spraying the organosilane solution at ambient conditions and letting it dry in a desiccator at 36 Torr for 20 min [6]. The open mesh serpentine layout of the EPEDs was designed using Adobe Illustrator CC (Adobe Systems Inc., San Jose, CA, USA) according to geometries previously reported in [10,11]. The minimum line width of the serpentine layout was kept at 200 μm (Figures 1b and 2b), the minimum resolution of our programmable razor printer (Silhouette Cameo^{TM}, Silhouette America Inc., Lindon, UT, USA), which uses a 100-μm-thick blade as the cutting tool. Prior to shaping the serpentine layout of the EPEDs, we attached adhesive copper tape (for copper-based EPEDs, Figure 1) or stencil printed Ag/AgCl ink (for Ag/AgCl-based EPEDs) on the functionalized paper. These functionalized paper substrates covered with a conductive layer (thickness of the composite ranging 70–190 μm) were then attached to a water-soluble tape (Aquasol Corp., North Tonawanda, NY, USA), which acted as the transfer layer to mount the EPEDs on skin (Figure 1c–f). Prior to the placement of EPEDs on skin, we sprayed medical glue (Medique products, Fort Myers, FL, USA) over the skin to maintain the conformal contact of the EPEDs on stretching. The transfer layer was then dissolved under a stream of running water (Figure 1e,f).

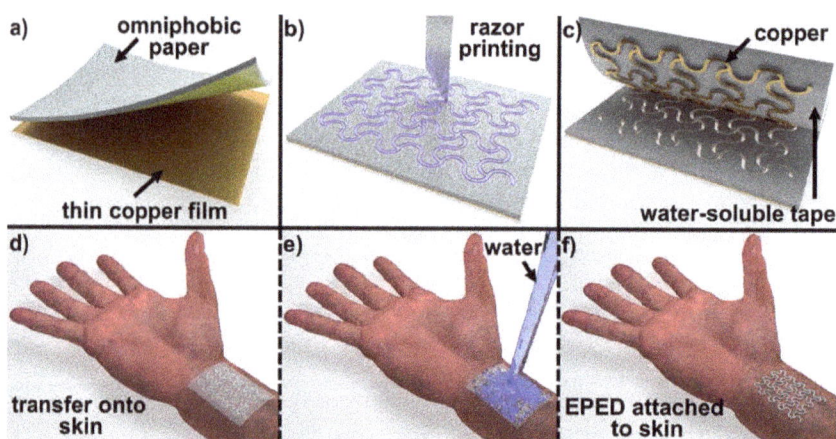

Figure 1. Fabrication of epidermal paper-based electronic devices (EPEDs) using razor printing: (**a**) A layer of omniphobic paper is glued to a thin metallic film that serves as a conductive layer (alternatively Ag/AgCl ink can be directly deposited on omniphobic paper); (**b**) A 100-μm-thick razor blade shapes the ensemble into a serpentine pattern; (**c**) A water-soluble tape, attached to the paper side of the EPED, is used as a temporary substrate for transfer onto skin; (**d**) The EPED is transferred onto skin previously sprayed with medical glue; (**e**) Placing the EPED under a stream of running water dissolves the temporary substrate; (**f**) EPED conformally attached to the skin.

2.3. Physiological Signal Measurement with EPEDs

We recorded ECG, EMG, and EOG signals using a three-electrode configuration [7]. The physiological signals were amplified, filtered, and displayed using a commercial electrophysiological recorder (Backyard Brains, Ann Arbor, MI, USA) coupled to a portable open-source microcontroller (UNO, Arduino Inc.). The thickness of the medical glue layer deposited on the skin is <2 µm [9], minimally affecting the performance of the EPED while recording physiological signals.

We attached external cables (28 AWG) directly onto the conductive layer of the EPEDs (over the contact pad area) using a small amount of low melting point soldering paste (SMD291AX, Chip Quik Inc., Niagara Falls, NY, USA). To perform underwater experiments, an extra layer of medical glue was deposited over the skin to encapsulate the flat connection between the EPEDs and the cables. Any excess of medical glue sprayed on the skin accumulated along the lateral walls of the EPED, preventing those exposed areas of the conductive layer from short-circuiting while under water.

To compare the performance of EPEDs with conventional electrodes, we ran parallel experiments using EPEDs and commercially available foam electrodes (Medline Industries Inc., Northfield, IL, USA). We coated the surface of the foam electrodes in contact with the skin with a conductive electrode gel (SPECTRA® 360, Parker Laboratories Inc., Fairfield, NJ, USA) to ensure a good electrical contact.

2.4. Characterization of Wirelessly Powered EPEDs

To wirelessly power functional components, such as LEDs, we attached a miniaturized half-wave rectifier fabricated using SMD components (Table S3) to the EPED antennas. We studied the wireless power transfer capabilities of EPEDs by performing a frequency-dependent characterization using a vector network analyzer (E5071B ENA, Agilent Technologies, Santa Clara, CA, USA). We used a copper coil (18 AWG wire, 6 turns, 5 cm diameter) connected to the network analyzer through an SMA connector (Digi-Key Electronics, Thief River Falls, MN, USA) to transfer wireless power to the EPEDs. All EPEDs were characterized passively at a distance of 15 cm from the center of the coil, in an orientation perpendicular to the axis of the coil. The network analyzer was programmed to record the real and imaginary parts of the impedance at 1601 frequency points linearly spaced in the range 1–20 MHz, finding the resonant frequency of the EPED using the min-phase method [39]. To enable the wireless powering of EPEDs, the coil was excited at the resonant frequency with a sinusoidal signal generated by a waveform generator (DG4062 Series, RIGOL Technologies Inc., Beaverton, OR, USA).

2.5. Heat Therapy Using EPEDs

We used EPEDs with copper and Ag/AgCl ink as the conductive layers to apply heat uniformly to the skin of the user. The thermal distribution created by the EPEDs was imaged using an infrared (IR) camera (FLIR E8, Wilsonville, OR, USA). We used a DC power supply (DP832A, RIGOL Technologies Inc., Beaverton, OR, USA) to generate heat through the resistive EPED, applying power levels below FCC guidelines (<2 W). To ensure the accuracy of the real-time monitoring of the temperature of the skin, we kept the distance between the IR camera and the EPED fixed at 20 cm during all the experiments.

2.6. Scanning Electron Microscopy (SEM)

We used a scanning electron microscope (Nova NanoSEM 200, FEI, Hillsboro, OR, USA) to examine the structure of the fabricated EPEDs. Before imaging, we used a sputter coater (208HR, Cressington, UK) to create a uniform conductive coating of ~10 nm platinum, using a D.C. current of 40 mA for 60 s. SEM images of the samples were captured at an electron accelerating potential of 5 kV, spot size 3, and working distance of 5 mm using an Everhart-Thornley detector (ETD).

2.7. Mechanical Characterization of EPEDs

We obtained stress-strain characteristics of bare paper substrates as well as fabricated EPEDs using a universal testing machine (MTS insight 10, MTS Systems Corp., Eden Prairie, MN, USA) equipped with a 100 N load cell (model 661.18.F01) according to ASTM D828-16 specifications. For the bare paper substrates, we fixed the gage length at 50 mm and applied a loading rate of 10 mm/min; while for the EPEDs, we had a gage length of 10 mm (comparable to the size of the device) and a loading rate of 5 mm/min.

3. Results and Discussion

3.1. Working Principle of EPEDs

Figure 2a shows the two layers of the EPEDs fabricated using razor printing (fabrication steps detailed in Figure 1): a conductive 20-μm-thick copper film in contact with the skin of the user and a silanized paper support (thickness ranging from 70 to 180 μm) that exhibits a static contact angle of 156°. The silane used to render paper omniphobic (both hydrophobic and oleophobic) prevents the EPEDs from being wetted by aqueous solutions and organic liquids with surface tension as low as 28 mN m^{-1} [6]. The covalent bonds generated between the alkyl trichlorosilanes and the cellulose fibers of the paper during the functionalization process are stable both in ambient conditions and under water for temperatures up to 150 °C [34]. Moreover, the chemical modification of the cellulose fibers of the paper do not affect its porosity (Figure 2c inset), preserving the gas permeability of the paper. The razor printing method used to fabricate the EPED enables the fabrication of flexible electrodes with a linewidth of 200 μm (Figure 2b) and a thickness of 78 μm when using Ag/AgCl ink as the conductive layer (70 μm is the thickness of the paper and 8 μm is the average thickness of the Ag/AgCl ink, see Figure 2c). The low thickness of the EPEDs fabricated using razor printing ensure their conformability to skin even when it wrinkles due to compression forces (Figure 2d) [15]. After dissolving the water-soluble transfer layer, 70-μm-thick EPEDs adhere to the skin solely by van der Waals and capillary forces, without requiring the spray-on medical glue. However, for a more robust adhesion of the EPEDs, we sprayed the skin with medical glue in all cases, regardless of the thickness of the paper used as a substrate. Since the thickness of the sprayed layer of glue is very small (<2 μm; [9]), its use does not adversely affect the functionality of the EPEDs or increase experimental noise in any significant way. The solvent of the medical glue sprayed on the skin of the user prior to placing the EPED does not affect the wetting properties of the paper substrate, which remains omniphobic after the medical glue dries. The omniphobic cellulose fibers provide a mechanical reinforcement to the thin film metals and conductive inks used in the conductive layer of the EPED, allowing them to withstand accidental stresses up to 2.5 MPa without tearing (Figure 2e). Despite the limited stretchability of unpatterned paper (~4%), the serpentine pattern used in the design of the EPED electrodes, enables these epidermal devices to endure stretching up to ~58% before failure. As a comparison, the maximum strain of human skin is ~30% [40].

3.2. Realtime Monitoring of Cardiac Activity

We recorded ECG signals from a human subject by attaching copper-based EPED electrodes on the wrist (measurement and ground) and the back of the hand (reference), as shown in Figure 3a. Figure 3b (top) shows the ECG signals recorded with the EPED electrodes. The silanization of the paper substrate to render EPEDs omniphobic allows us to capture high quality ECG signals even with the EPEDs completely immersed in water (Figure 3b bottom). We compared the signal to noise ratio (SNR) of EPED electrodes to conventional foam electrodes by placing them on the same locations of the hand (Figure 3c top). ECG measurements acquired by razor printed EPED electrodes (SNR$_{ECG-EPED,air}$ = 12.20 dB, SNR$_{ECG-EPED,water}$ = 10.37 dB; Table S2) exhibit no significant difference from conventional foam electrodes in air (Figure 3c bottom, SNR$_{ECG-foam,air}$ = 11.28 dB).

Conventional electrodes, though, cannot reliably capture ECG signals under water due to the swelling of their hydrogel terminals and their subsequent delamination and short-circuit.

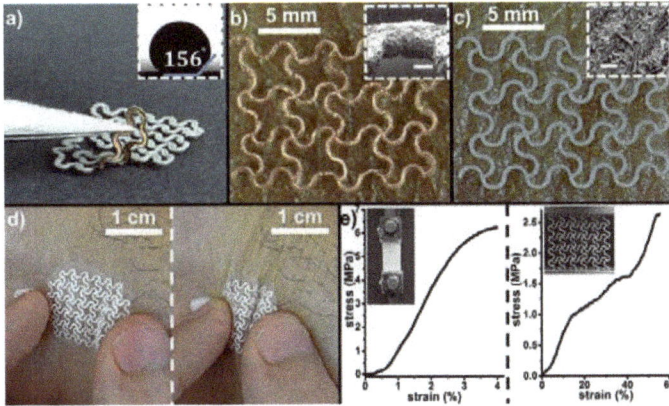

Figure 2. Razor printed EPEDs: (**a**) Omniphobic EPEDs comprising a thin conducting layer and a silanized paper serving as a back support. Inset shows the apparent contact angle of a 10 µL water droplet on top of the silanized paper substrate; (**b**) EPED on top of skin, with 20-µm-thick copper film as a conducting layer (copper facing up). The inset shows an SEM image of the 70-µm-thick patterned paper substrate (scale bar is 50 µm); (**c**) EPED with an 8-µm-thick layer of Ag/AgCl ink deposited on top of the omniphobic paper (Ag/AgCl facing up). The inset shows an SEM image of the Ag/AgCl/paper electrode, demonstrating that neither the functionalization of the paper nor the subsequent deposition of Ag/AgCl ink clogged the porous structure of the paper substrate (scale bar is 100 µm); (**d**) Conforming of EPEDs to skin bending and buckling due to severe compression; (**e**) Left: Representative stress-strain curve of an unpatterned paper substrate. Inset shows the experimental set up used. Right: Stress-strain curve of an Ag/AgCl EPED showing how the razor patterning of the EPED improves its stretching when compared to unpatterned paper. Inset shows the mechanical characterization of a representative EPED sample.

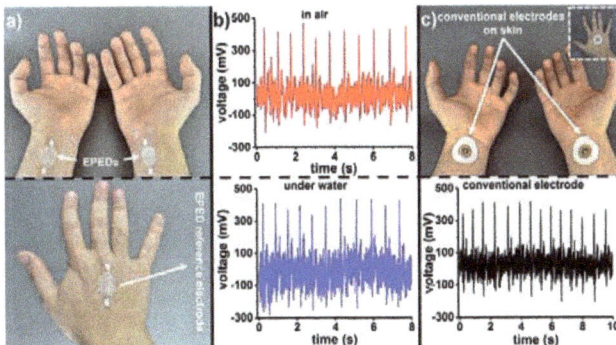

Figure 3. Comparison between the performance of razor printed copper-based EPEDs and conventional foam electrodes to record ECG signals: (**a**) Top: EPED measurement and ground electrodes used to record ECG signals from the wrist of a subject. Bottom: Reference EPED electrode; (**b**) Top: ECG signals recorded in air using razor printed EPEDs. Bottom: ECG signals recorded with both hands under water; (**c**) Top: Conventional foam electrodes placed at the same locations of the wrist as (**a**). The inset shows the location of the reference foam electrode on the back of the hand. Bottom: ECG signals recorded in air using conventional foam electrodes.

3.3. Real-Time Monitoring of Muscle Activity

We used copper-based EPEDs to record EMG signals from the forearm by placing the measurement and ground electrodes along the flexor muscle and the reference electrode on the back of the hand (Figure 4a). Figure 4b (top) shows the EMG signals recorded with EPEDs while lifting a 4.5 kg dumbbell, holding it for 5 s, and resting for 10 s. The omniphobic character of the EPED provided by the silanization of the paper substrate, enables EMG signals to be captured in high moisture environments without significant experimental noise or short circuiting the measuring electrodes. To demonstrate the moisture-independent collection of EMG signals, we repeated the measurements while keeping the arm in a water bath. We observed no significant difference in performance between the razor printed EPEDs ($SNR_{EMG-EPED,air}$ = 31.79 dB, $SNR_{EMG-EPED,water}$ = 30.16 dB; Table S2) and conventional foam electrodes ($SNR_{EMG-foam,air}$ = 26.58 dB) to record EMG signals in air. Conventional electrodes, however, are not capable to record EMG signals under water due to the short-circuit of their terminals.

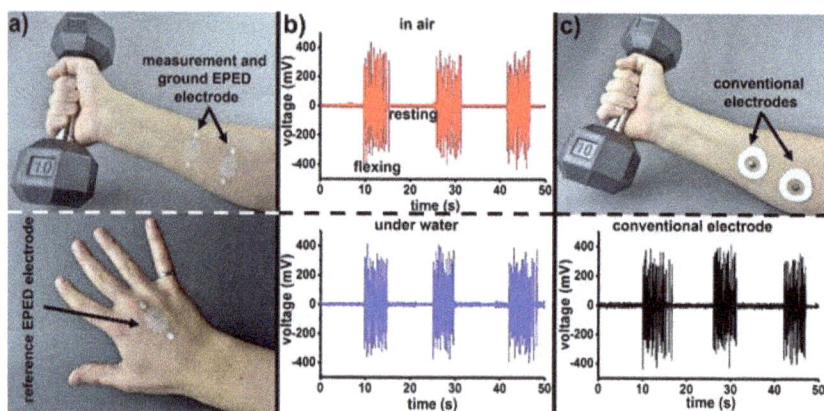

Figure 4. Comparison between the performance of razor printed EPEDs and conventional foam electrodes to monitor EMG signals: (**a**) Top: EPED measurement and ground electrodes used to record EMG signals from the forearm of an exercising subject. Bottom: Reference EPED electrode; (**b**) Top: EMG signals recorded in air using razor printed EPEDs. Bottom: EMG signals recorded under water; (**c**) Top: conventional foam electrodes placed at the same locations of the forearm as (**a**). Bottom: EMG signals recorded in air using conventional foam electrodes.

3.4. Monitoring Eye Motion

We used three copper-based EPED electrodes placed on the cheekbone (ground), forehead (measurement), and neck (reference) to capture EOG signals and to monitor the movement of the eye (Figure 5a). The mechanical conformability of EPEDs enabled the identification of the movement of the eyes (up and down) as well as blinking events with minimal experimental noise ($SNR_{EOG-EPED,air}$ = 33.77 dB; Figure 5b, Table S2). When compared with conventional foam electrodes ($SNR_{EOG-foam,air}$ = 31.17 dB), EPEDs exhibit better performance upon the natural moves of the user (Figure 5c).

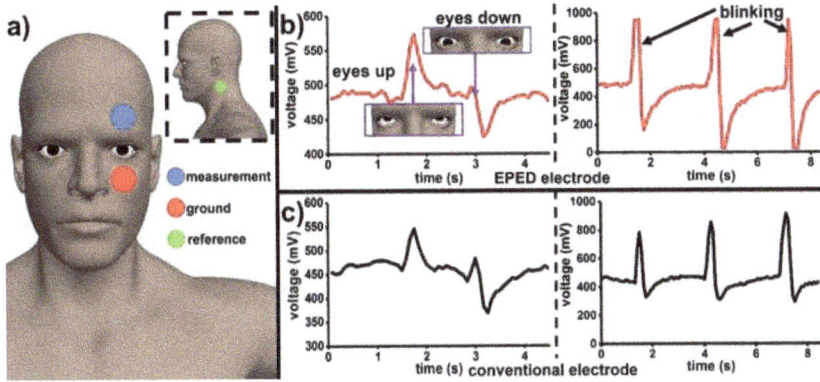

Figure 5. Monitoring eye motion using razor printed EPEDs: (**a**) Location of the measuring, ground, and reference copper-based EPED electrodes; (**b**) Left: EOG signal identifying eye movements (up and down) using EPEDs; Right: Identification of blinking events using EPEDs; (**c**) Left: Identification of eye movements (up and down) using conventional foam electrodes located as shown in (**a**); Right: Identification of blinking events using conventional foam electrodes.

3.5. Wireless Powering of EPEDs

The low resistivity (~20 nΩ m for copper-based EPEDs) of EPED antennas make them suitable for wireless power transfer based applications (Figure 6). Figure 6a,b shows a 10 mm EPED antenna with an LED and a rectifier circuit mounted on skin, being powered using far-field electromagnetic waves emitted from a primary coil placed 15 cm away. The geometry of the EPED antenna was chosen to match previously reported wireless epidermal stimulators [11]. This square pancake coil has only three loops to minimize their shaping with the razor printer, since we experimentally found that coils with three loops were able to efficiently power the LED wirelessly via inductive coupling at a distance of 15 cm. The line width of the antenna was made to match the minimum resolution of our razor printer (~200 μm). After the coil was shaped with the razor printer, we folded the external end of the coil towards the center to make both ends of the coil to rest flat at a distance of 2 mm and soldered the SMD components (Table S3) between the ends of the coil (Figure 6a,b). We analyzed the frequency-dependent electrical characteristics of this EPED using methods described in Section 2.4 (Figure 6c). We recorded the real and imaginary components of the impedance, $Z = R + jX$, $|Z| = (R^2 + X^2)^{1/2}$, where the real part, R, is the resistance, and the imaginary part, X, is the reactance. We used the recorded components of the impedance as a function of frequency to calculate electrical characteristics of the EPEDs such as inductance $L = X/2\pi f$, phase $\theta = \tan^{-1}(X/R)$, and quality factor $Q = X/R$. The resonant frequency f_0 of the EPED is determined by the min-phase method [39,41]: the frequency at which the θ response is minimized is taken as the resonant frequency of the EPED when coupled with the primary coil. Since the primary coil is connected to the network analyzer for a one-port measurement, only the S_{11} parameter is recorded (Figure 6d). The wireless power transfer efficiency is calculated as $\eta = (1 - |S_{11}|^2) \times 100\%$, where $|S_{11}|^2$ is defined as the reflectance. We found that the omniphobic functionalization of the paper does not significantly modify the resonant frequency or the power transfer efficiency of the EPED (Figure 6d,e). Modifying the values of the capacitors used to rectify the wireless signal, the wireless power transfer efficiency of the EPED can be easily optimized for a given frequency following the impedance-matching optimization method for magnetic resonance coupling systems [42].

Multiple EPEDs placed in close proximity can be selectively powered, if these EPEDs have different resonant frequency peaks (due to their different geometry or rectifying circuit). Figure 7 summarizes the passive electrical characteristics of a system of 2 EPEDs, one 8 mm and another 10 mm, placed side by side, but not in contact with each other. The different resonant frequencies of the EPEDs (9.0 MHz for 8 mm side EPED; 10.4 MHz for 10 mm side EPED) enable their selective activation, independently or at the same time, depending on the desired application (Figure 7d).

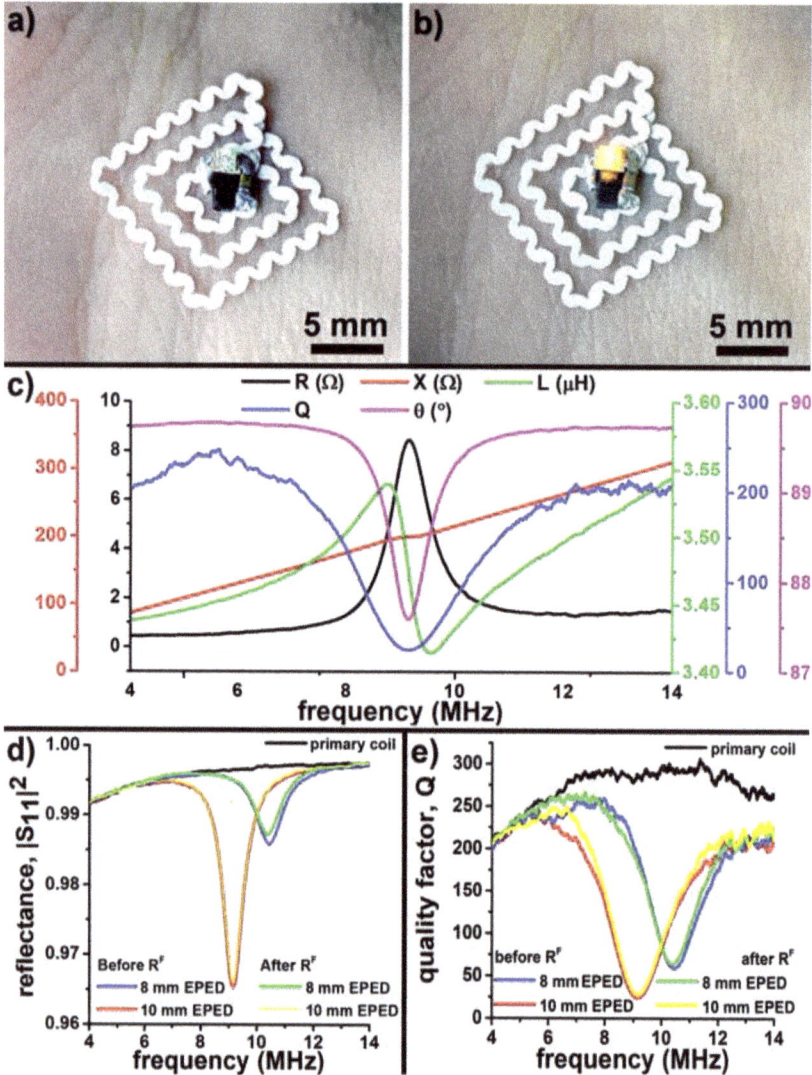

Figure 6. Electrical characteristics of wirelessly powered copper-based EPEDs: (**a**) Square EPED antenna (10 mm side) coupled with an SMD rectifier-LED circuit and attached to the skin of the wrist; (**b**) The LED is wirelessly powered using a primary coil 15 cm away (not shown in picture) running alternating currents at a resonant frequency of 9.0 MHz; (**c**) Frequency-dependent passive characteristics (Resistance, R, Reactance, X, Phase, θ, Inductance, L, and Quality factor, Q) of the EPED shown in (**a,b**). The frequency at which the phase θ is minimum is the resonant frequency of the EPED; (**d,e**) Effect of EPED size and silanization on the wireless power transfer efficiency ($\eta = 1 - |S_{11}|^2$) and the quality factor (Q). The silanization process has a negligible effect on η and Q. η and Q decrease when the size of the EPED is reduced.

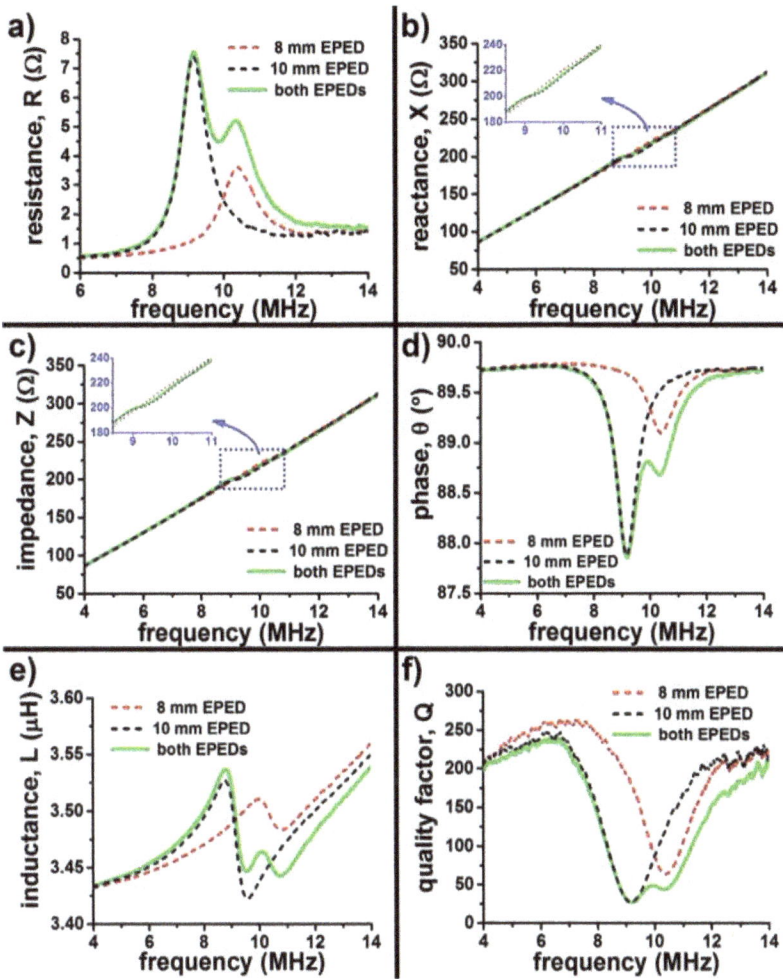

Figure 7. Selective powering of multiple EPEDs by varying excitation frequency, using the same primary coil. The green solid curves in all panels correspond to the frequency dependent passive characteristics when both EPEDs (8 mm and 10 mm side) are 15 cm away from the primary coil. The dashed lines represent individual characteristics: (**a**) Resistance, R; (**b**) Reactance, X; (**c**) Impedance, Z; (**d**) Phase, θ; (**e**) Inductance, L; (**f**) Quality factor, Q.

3.6. Localized Heat Therapy

Epidermal heat therapy is commonly employed in cancer treatments [43] and in orthopedics for alleviating joint pain [44]. Figure 8a shows a copper-based EPED fabricated to apply localized heat therapy on the skin. Heat is produced in accordance with Joule's law of heating by running D.C. power through the EPED. The serpentine layout of this EPED was designed according to previously reported epidermal electronic devices [10,11] and razor printed with a minimum linewidth of 200 μm. We monitored the temperature distribution produced by the EPED using an IR camera, limiting the maximum temperature applied to the skin to 42 °C (Figure 8b). Each of the quadrants of the EPED has two independent contact pads that enable their individual activation to provide localized doses of heat (Figure 8b inset). Figure 8c shows the time-dependence of the heating process for different

D.C. powers. The low specific heat of copper ensures that the EPED temperature rapidly stabilizes as power is applied, and restores quickly to room temperature once the power supply is turned off. The omniphobic properties of the EPED remain unaffected after the heating cycles.

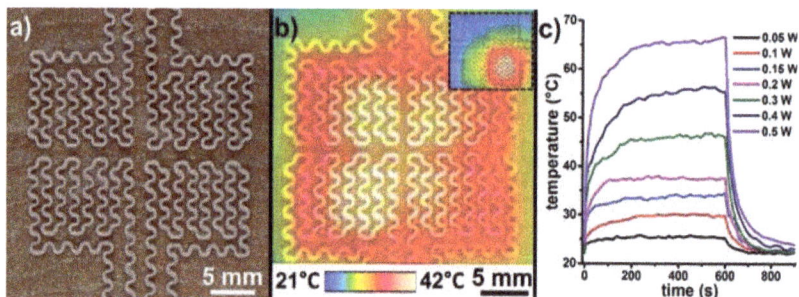

Figure 8. Application of localized heat therapy using razor printed EPEDs: (a) Copper-based thermotherapy EPED mounted on skin; (b) IR image of the EPED shown in (a) during thermotherapy (inset shows the application of localized heat by the selective activation of only one quadrant of the EPED); (c) Temperature-time response of the thermotherapy EPED for different D.C. powers.

3.7. Thermal Sensing

The temperature dependence of the resistivity of the conductive layer of the EPEDs enables their use as wearable thermometers (Figure 9). The linear relationship between the increment in resistance of the EPED and its temperature allowed us to calculate the sensitivity of Ag/AgCl- and copper-based EPED thermometers (Figure 9a). The sensitivity of Ag/AgCl- and copper-based EPEDs are 0.01 Ω/°C and 0.001 Ω/°C, respectively. We characterized the time response of the Ag/AgCl EPEDs by placing them over a surface at room temperature (t = 0 s) and placing an aluminum cylinder preheated to different reference temperatures (t = 10 s). We observed that Ag/AgCl EPEDs required less than 1 s to reach reference temperatures in the clinically relevant range (Figure 9b).

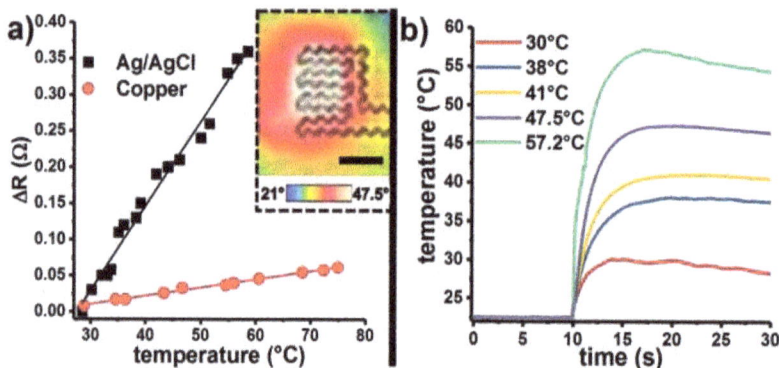

Figure 9. Sensing temperature using EPEDs: (a) Change of the EPED resistance as a function of temperature for EPEDs with a thin copper film (solid red dots) and Ag/AgCl ink (solid black squares) as conductive layers. The inset shows an IR image of an Ag/AgCl-based EPED when a small aluminum cylinder at 47.5 °C is placed on its surface for 10 s and then removed. Scale bar is 5 mm; (b) Response of the Ag/AgCl-based EPED thermometers when an aluminum cylinder at different temperatures is placed in contact with the EPED at t = 10 s.

4. Conclusions

This work reports the simple, inexpensive, and scalable, fabrication of epidermal paper-based electronic devices (EPEDs) using a bench-top razor printer. EPEDs fabricated using silanized paper can be used as moisture-insensitive epidermal electrodes, with a cost so low that it makes them compatible with single-use applications (see Table S1). Razor printed EPEDs fabricated using copper film or Ag/AgCl ink are easy to mount on skin, conforming to its natural moves, and exhibit good mechanical contact with the user and a stable electrical performance upon stretching. Copper-based EPEDs exhibit low resistivity values (~20 nΩ m), enabling their use as efficient electrophysiological monitors, thermotherapeutic devices, and wirelessly powered systems. The low resistance of copper-based EPEDs, however, makes it difficult to detect changes in the resistance caused by environmental temperature. Ag/AgCl-based EPEDs have higher resistivity values (~110 nΩ m), facilitating their use as temperature sensors since small changes in the environmental temperature induce larger changes in the resistance of the devices. The fibrous structure of the paper substrates of the EPEDs makes them breathable when their conductive layer is porous, such as Ag/AgCl-based EPEDs. The adhesion of a continuous copper film to the paper, however, compromises the passage of gases across the EPED. We demonstrated the omniphobic character of razor printed EPEDs by efficiently recording ECGs, EMGs, and EOGs in air and under water without any significant decrease in performance. The use of razor printing to fabricate EPEDs, at its present level of development, also has two limitations: (i) The minimum line width of the serpentine traces is 200 µm; (ii) The shear forces applied during high-speed cutting processes can lead to the delamination of the conductive layer from the omniphobic paper support if the adhesive used to secure both layers is not properly chosen. The wide range of adhesive materials and films compatible with razor printing, however, can ameliorate this limitation. We expect that the proposed method to fabricate inexpensive wearable electrodes will facilitate the adoption of epidermal electronics in personalized medicine, especially in resource-limited and home environments.

Supplementary Materials: The following are available online at http://www.mdpi.com/2072-666X/9/9/420/s1, Table S1: Itemized cost per device of each of the components integrating a razor printed EPED; Table S2: Signal-to-Noise Ratio (SNR) for electrophysiological signals recorded with EPEDs and conventional foam electrodes; Table S3: Layout of half-wave rectifying circuit connected to EPED antenna.

Author Contributions: R.V.M and B.S. conceived the research and designed the experiments. B.S. fabricated the EPEDs and performed the experiments related to physiological signal monitoring, thermal stimulation, and mechanical characterization. D.G. performed electrical and mechanical characterization experiments as well as the structural characterization of the devices using electron microscopy. B.S., D.G. and R.V.M. co-wrote the paper.

Acknowledgments: The authors gratefully acknowledge start-up funding from Purdue University. B.S. also acknowledges the funding from Procter & Gamble (grant no. 209621). D.G. thanks the Ross Fellowship program at Purdue University for providing partial support of his work.

Conflicts of Interest: The authors declare no conflict of interest.

References

1. Hayward, J.; Chansin, G.; Zervos, H. *Wearable Technology 2017–2027: Markets, Players, Forecasts*; IDTechEx: Santa Clara, CA, USA, 2017.

2. Lobodzinski, S.S.; Laks, M.M. New devices for very long-term ECG monitoring. *Cardiol. J.* **2012**, *19*, 210–214. [CrossRef] [PubMed]

3. Araki, H.; Kim, J.; Zhang, S.; Banks, A.; Crawford, K.E.; Sheng, X.; Gutruf, P.; Shi, Y.; Pielak, R.M.; Rogers, J.A. Materials and Device Designs for an Epidermal UV Colorimetric Dosimeter with Near Field Communication Capabilities. *Adv. Funct. Mater.* **2017**, *27*, 1604465. [CrossRef]

4. Choi, H.; Luzio, S.; Beutler, J.; Porch, A. Microwave noninvasive blood glucose monitoring sensor: Human clinical trial results. In Proceedings of the 2017 IEEE MTT-S International Microwave Symposium (IMS), Honolulu, HI, USA, 4–9 June 2017; pp. 876–879.

5. Lee, H.; Song, C.; Hong, Y.S.; Kim, M.S.; Cho, H.R.; Kang, T.; Shin, K.; Choi, S.H.; Hyeon, T.; Kim, D.-H. Wearable/disposable sweat-based glucose monitoring device with multistage transdermal drug delivery module. *Sci. Adv.* **2017**, *3*, e1601314. [CrossRef] [PubMed]

6. Glavan, A.C.; Martinez, R.V.; Maxwell, E.J.; Subramaniam, A.B.; Nunes, R.M.D.; Soh, S.; Whitesides, G.M. Rapid fabrication of pressure-driven open-channel microfluidic devices in omniphobic R^F paper. *Lab Chip* **2013**, *13*, 2922–2930. [CrossRef] [PubMed]

7. Jeong, J.-W.; Kim, M.K.; Cheng, H.; Yeo, W.-H.; Huang, X.; Liu, Y.; Zhang, Y.; Huang, Y.; Rogers, J.A. Capacitive Epidermal Electronics for Electrically Safe, Long-Term Electrophysiological Measurements. *Adv. Healthc. Mater.* **2014**, *3*, 642–648. [CrossRef] [PubMed]

8. Park, S.I.; Brenner, D.S.; Shin, G.; Morgan, C.D.; Copits, B.A.; Chung, H.U.; Pullen, M.Y.; Noh, K.N.; Davidson, S.; Oh, S.J.; et al. Soft, stretchable, fully implantable miniaturized optoelectronic systems for wireless optogenetics. *Nat. Biotechnol.* **2015**, *33*, 1280–1286. [CrossRef] [PubMed]

9. Yeo, W.-H.; Kim, Y.-S.; Lee, J.; Ameen, A.; Shi, L.; Li, M.; Wang, S.; Ma, R.; Jin, S.H.; Kang, Z.; et al. Multifunctional Epidermal Electronics Printed Directly onto the Skin. *Adv. Mater.* **2013**, *25*, 2773–2778. [CrossRef] [PubMed]

10. Zhang, Y.; Wang, S.; Li, X.; Fan, J.A.; Xu, S.; Song, Y.M.; Choi, K.-J.; Yeo, W.-H.; Lee, W.; Nazaar, S.N.; et al. Experimental and Theoretical Studies of Serpentine Microstructures Bonded To Prestrained Elastomers for Stretchable Electronics. *Adv. Funct. Mater.* **2014**, *24*, 2028–2037. [CrossRef]

11. Fan, J.A.; Yeo, W.-H.; Su, Y.; Hattori, Y.; Lee, W.; Jung, S.-Y.; Zhang, Y.; Liu, Z.; Cheng, H.; Falgout, L.; et al. Fractal design concepts for stretchable electronics. *Nat. Commun.* **2014**, *5*, 3266. [CrossRef] [PubMed]

12. Efimenko, K.; Rackaitis, M.; Manias, E.; Vaziri, A.; Mahadevan, L.; Genzer, J. Nested self-similar wrinkling patterns in skins. *Nat. Mater.* **2005**, *4*, 293–297. [CrossRef] [PubMed]

13. Webb, R.C.; Bonifas, A.P.; Behnaz, A.; Zhang, Y.; Yu, K.J.; Cheng, H.; Shi, M.; Bian, Z.; Liu, Z.; Kim, Y.-S.; et al. Ultrathin conformal devices for precise and continuous thermal characterization of human skin. *Nat. Mater.* **2013**, *12*, 938–944. [CrossRef] [PubMed]

14. Kim, J.; Banks, A.; Cheng, H.; Xie, Z.; Xu, S.; Jang, K.-I.; Lee, J.W.; Liu, Z.; Gutruf, P.; Huang, X.; et al. Epidermal Electronics with Advanced Capabilities in Near-Field Communication. *Small* **2015**, *11*, 906–912. [CrossRef] [PubMed]

15. Liu, Y.; Pharr, M.; Salvatore, G.A. Lab-on-Skin: A Review of Flexible and Stretchable Electronics for Wearable Health Monitoring. *ACS Nano* **2017**, *11*, 9614–9635. [CrossRef] [PubMed]

16. Tobjörk, D.; Österbacka, R. Paper Electronics. *Adv. Mater.* **2011**, *23*, 1935–1961. [CrossRef] [PubMed]

17. Russo, A.; Ahn, B.Y.; Adams, J.J.; Duoss, E.B.; Bernhard, J.T.; Lewis, J.A. Pen-on-Paper Flexible Electronics. *Adv. Mater.* **2011**, *23*, 3426–3430. [CrossRef] [PubMed]

18. Lessing, J.; Glavan, A.C.; Walker, S.B.; Keplinger, C.; Lewis, J.A.; Whitesides, G.M. Inkjet Printing of Conductive Inks with High Lateral Resolution on Omniphobic "R^F Paper" for Paper-Based Electronics and MEMS. *Adv. Mater.* **2014**, *26*, 4677–4682. [CrossRef] [PubMed]

19. Bollström, R.; Määttänen, A.; Tobjörk, D.; Ihalainen, P.; Kaihovirta, N.; Österbacka, R.; Peltonen, J.; Toivakka, M. A multilayer coated fiber-based substrate suitable for printed functionality. *Org. Electron.* **2009**, *10*, 1020–1023. [CrossRef]

20. Cui, L.-F.; Ruffo, R.; Chan, C.K.; Peng, H.; Cui, Y. Crystalline-Amorphous Core–Shell Silicon Nanowires for High Capacity and High Current Battery Electrodes. *Nano Lett.* **2009**, *9*, 491–495. [CrossRef] [PubMed]

21. Hamedi, M.M.; Ainla, A.; Güder, F.; Christodouleas, D.C.; Fernández-Abedul, M.T.; Whitesides, G.M. Integrating Electronics and Microfluidics on Paper. *Adv. Mater.* **2016**, *28*, 5054–5063. [CrossRef] [PubMed]

22. Calvert, P. Inkjet Printing for Materials and Devices. *Chem. Mater.* **2001**, *13*, 3299–3305. [CrossRef]

23. Kim, D.-H.; Kim, Y.-S.; Wu, J.; Liu, Z.; Song, J.; Kim, H.-S.; Huang, Y.Y.; Hwang, K.-C.; Rogers, J.A. Ultrathin Silicon Circuits With Strain-Isolation Layers and Mesh Layouts for High-Performance Electronics on Fabric, Vinyl, Leather, and Paper. *Adv. Mater.* **2009**, *21*, 3703–3707. [CrossRef]

24. Hyun, W.J.; Secor, E.B.; Hersam, M.C.; Frisbie, C.D.; Francis, L.F. High-Resolution Patterning of Graphene by Screen Printing with a Silicon Stencil for Highly Flexible Printed Electronics. *Adv. Mater.* **2015**, *27*, 109–115. [CrossRef] [PubMed]

25. Huang, J.; Zhu, H.; Chen, Y.; Preston, C.; Rohrbach, K.; Cumings, J.; Hu, L. Highly Transparent and Flexible Nanopaper Transistors. *ACS Nano* **2013**, *7*, 2106–2113. [CrossRef] [PubMed]

26. Liu, H.; Crooks, R.M. Paper-Based Electrochemical Sensing Platform with Integral Battery and Electrochromic Read-Out. *Anal. Chem.* **2012**, *84*, 2528–2532. [CrossRef] [PubMed]
27. Hu, L.; Zheng, G.; Yao, J.; Liu, N.; Weil, B.; Eskilsson, M.; Karabulut, E.; Ruan, Z.; Fan, S.; Bloking, J.T.; et al. Transparent and conductive paper from nanocellulose fibers. *Energy Environ. Sci.* **2013**, *6*, 513–518. [CrossRef]
28. Zhao, R.; Zhang, X.; Xu, J.; Yang, Y.; He, G. Flexible paper-based solid state ionic diodes. *RSC Adv.* **2013**, *3*, 23178–23183. [CrossRef]
29. Pal, A.; Cuellar, H.E.; Kuang, R.; Caurin, H.F.N.; Goswami, D.; Martinez, R.V. Self-Powered, Paper-Based Electrochemical Devices for Sensitive Point-of-Care Testing. *Adv. Mater. Technol.* **2017**, *2*, 1700130. [CrossRef]
30. Anagnostou, D. *Organic Paper-Based Antennas*; WIT Press: Billerica, MA, USA, 2014.
31. Teisala, H.; Tuominen, M.; Kuusipalo, J. Superhydrophobic Coatings on Cellulose-Based Materials: Fabrication, Properties, and Applications. *Adv. Mater. Interfaces* **2014**, *1*, 1300026. [CrossRef]
32. Li, X.; Ballerini, D.R.; Shen, W. A perspective on paper-based microfluidics: Current status and future trends. *Biomicrofluidics* **2012**, *6*, 011301. [CrossRef] [PubMed]
33. Carrilho, E.; Martinez, A.W.; Whitesides, G.M. Understanding Wax Printing: A Simple Micropatterning Process for Paper-Based Microfluidics. *Anal. Chem.* **2009**, *81*, 7091–7095. [CrossRef] [PubMed]
34. Glavan, A.C.; Martinez, R.V.; Subramaniam, A.B.; Yoon, H.J.; Nunes, R.M.D.; Lange, H.; Thuo, M.M.; Whitesides, G.M. Omniphobic "R^F Paper" Produced by Silanization of Paper with Fluoroalkyltrichlorosilanes. *Adv. Funct. Mater.* **2014**, *24*, 60–70. [CrossRef]
35. Liu, X.; Mwangi, M.; Li, X.; O'Brien, M.; Whitesides, G.M. Paper-based piezoresistive MEMS sensors. *Lab Chip* **2011**, *11*, 2189–2196. [CrossRef] [PubMed]
36. Chitnis, G.; Ding, Z.; Chang, C.-L.; Savran, C.A.; Ziaie, B. Laser-treated hydrophobic paper: An inexpensive microfluidic platform. *Lab Chip* **2011**, *11*, 1161–1165. [CrossRef] [PubMed]
37. Glavan, A.C.; Christodouleas, D.C.; Mosadegh, B.; Yu, H.D.; Smith, B.S.; Lessing, J.; Fernández-Abedul, M.T.; Whitesides, G.M. Folding Analytical Devices for Electrochemical ELISA in Hydrophobic R^H Paper. *Anal. Chem.* **2014**, *86*, 11999–12007. [CrossRef] [PubMed]
38. Giokas, D.L.; Tsogas, G.Z.; Vlessidis, A.G. Programming Fluid Transport in Paper-Based Microfluidic Devices Using Razor-Crafted Open Channels. *Anal. Chem.* **2014**, *86*, 6202–6207. [CrossRef] [PubMed]
39. Huang, X.; Liu, Y.; Cheng, H.; Shin, W.-J.; Fan, J.A.; Liu, Z.; Lu, C.-J.; Kong, G.-W.; Chen, K.; Patnaik, D.; et al. Materials and Designs for Wireless Epidermal Sensors of Hydration and Strain. *Adv. Funct. Mater.* **2014**, *24*, 3846–3854. [CrossRef]
40. Maiti, R.; Gerhardt, L.-C.; Lee, Z.S.; Byers, R.A.; Woods, D.; Sanz-Herrera, J.A.; Franklin, S.E.; Lewis, R.; Matcher, S.J.; Carré, M.J. In vivo measurement of skin surface strain and sub-surface layer deformation induced by natural tissue stretching. *J. Mech. Behav. Biomed. Mater.* **2016**, *62*, 556–569. [CrossRef] [PubMed]
41. Sardini, E.; Serpelloni, M. Passive and Self-Powered Autonomous Sensors for Remote Measurements. *Sensors* **2009**, *9*, 943–960. [CrossRef] [PubMed]
42. Thackston, K.A.; Mei, H.; Irazoqui, P.P. Coupling Matrix Synthesis and Impedance-Matching Optimization Method for Magnetic Resonance Coupling Systems. *IEEE Trans. Microw. Theory Tech.* **2018**, *66*, 1536–1542. [CrossRef]
43. Schroeder, A.; Heller, D.A.; Winslow, M.M.; Dahlman, J.E.; Pratt, G.W.; Langer, R.; Jacks, T.; Anderson, D.G. Treating metastatic cancer with nanotechnology. *Nat. Rev. Cancer* **2012**, *12*, 39–50. [CrossRef] [PubMed]
44. Brosseau, L.; Yonge, K.; Welch, V.; Marchand, S.; Judd, M.; Wells, G.A.; Tugwell, P. Thermotherapy for treatment of osteoarthritis. *Cochrane Database Syst. Rev.* **2003**. [CrossRef] [PubMed]

micromachines

MDPI

Article

The Conformal Design of an Island-Bridge Structure on a Non-Developable Surface for Stretchable Electronics

Lin Xiao [1,2], Chen Zhu [1,2], Wennan Xiong [1,2], YongAn Huang [1,2,*] and Zhouping Yin [1,2]

[1] State Key Laboratory of Digital Manufacturing Equipment and Technology, Huazhong University of Science and Technology, Wuhan 430074, China; linxiao@hust.edu.cn (L.X.); zhuchen@hust.edu.cn (C.Z.); xiongwn@foxmail.com (W.X.); yinzhp@hust.edu.cn (Z.Y.)

[2] Flexible Electronics Research Center, Huazhong University of Science and Technology, Wuhan 430074, China

*   Correspondence: yahuang@hust.edu.cn

Received: 25 June 2018; Accepted: 2 August 2018; Published: 7 August 2018

Abstract: Conformal design of the island-bridge structure is the key to construct high-performance inorganic stretchable electronics that can be conformally transferred to non-developable surfaces. Former studies in conformal problems of epidermal electronics are mainly focused on soft surfaces that can adapt to the deformation of the electronics, which are not suitable for applications in hard, non-developable surfaces because of their loose surface constraints. In this paper, the conformal design problem for the island-bridge structure on a hard, non-developable surface was studied, including the critical size for island and stiffness and the demand for stretchability for the bridge. Firstly, the conformal model for an island on a part of torus surface was established to determine the relationship between the maximum size of the island and the curvatures of the surface. By combining the principle of energy minimization and the limit of material failure, a critical non-dimensional width for conformability was given for the island as a function of its thickness and interfacial adhesion energy, and the ratio of two principal curvatures of the surface. Then, the dependency of the tensile stiffness of the bridge on its geometric parameters was studied by finite element analysis (FEA) to guide the deterministic assembly of the islands on the surface. Finally, the location-dependent demands for the stretchability of the bridges were given by geometric mapping. This work will provide a design rule for stretchable electronics that fully conforms to the non-developable surface.

Keywords: island-bridge; conformal design; non-developable surface; stretchable electronics

1. Introduction

Stretchable electronics can be conformally transferred to various surfaces to perform multifunctional curvilinear electronics systems, such as electronic eye camera [1–3], 3D integumentary membranes [4,5], wearable devices [6–11], and smart aircraft skin [12,13]. The island-bridge structure is usually used in fabricating stretchable electronics, as it has made the most of high-performance, inorganic semiconductor materials. By placing intrinsic brittle materials on an unstretchable island to protect them from damage caused by strain, the whole device can suffer a large deformation without failure. When it is transferred to a hard, non-developable surface, strain will be produced in the device because of the geometric mismatch between the plane and non-developable surfaces, which may cause conformal problems for the device. On the one hand, although most of the strain is withstood by the bridge, strain still exists on the island. With the increase of the island size or the local curvatures of the surface, the strain on the island will increase as well and cause failure eventually. On the other hand, the strain in the device may cause the island to change position, which means stretchability is needed for the bridge to accommodate this change. Obviously, this demand for the stretchability of the bridges varies with the shape of the surface. Besides, the mismatch strain distribution is non-uniform, and it is dependent on the curvature distribution of the surface, which brings huge challenges in the deterministic assembly of the electronics. Considering that the island-bridge structure is a "mass-spring system" in the broad sense, the positon of the mass (island) in equilibrium can be decided once the stiffness of the spring (bridge) is known. So, it is possible to realize the deterministic assembly by predesigning the stiffness of the bridge. Hence, the conformal problems need to be studied to determine the critical size of the island, the demand for stretchability, and the stiffness of the bridge.

The conformal problems of the island have been studied in epidermal electronics [14–18]. However, the target surfaces of epidermal electronics are usually soft and can accommodate the deformation of the island by being stretched or bent. Regarding conformal problem of island on a hard, non-developable surface, only the island is under deformation, which brings new challenges for the design of island. Several researchers have studied the adhesion and buckling problem between the elastic plate and the rigid sphere using theoretical, experimental, and simulation methods [19–24]. Majidi et al. [19] have given a critical conformal width for circular and rectangular elastic plates using the principle of energy minimization. However, the limits of material failure have not been taken into consideration, so it may not be suited to electronic design. Besides, the former studies are based on a sphere, which produces great limitations on the use of these theories. Mitchell et al. [25] show that a sheet that conforms to a cap and a saddle will produce different strain responses, respectively. Hence, a theory based on a more common surface needs to be proposed eagerly.

The theoretical works for the design of the bridge are quite mature, and many researchers have made significant contributions to this field [26–35]. Current works in bridge design mainly aim to promote its stretchability; the works for solving demand for stretchability are very rare. Nevertheless, some sacrifices are usually needed in other aspects of the device to obtain higher stretchability, such as functional duty ratio and material choice, which may cause an additional performance loss in the device. So, appropriate stretchability for the bridge is needed to be designed according to actual demand. On the other hand, the theoretical solutions for the stiffness of the bridges are mainly for thick bridges because of the complicated post-buckling behaviors in thin bridges [27,34]. Yihui Zhang [31] and Wentao Dong [32] have studied the thin bridge using finite element analysis (FEA), given its stretchability, but the relationships between stiffness and its geometric parameters for thin bridges are still ungiven.

In the present study, the conformal behavior of the island and design demand for the bridge are studied. The layout of the paper is as follows. A mechanical model of the island on a part of torus surface is presented in Section 2, and a non-dimensional critical conformal width is given by the combination of the principle of energy minimization and the limits for material failure. Furthermore, an adhesion experiment for island is implemented to verify the validity of the theory. Section 3 describes the relationship between the tensile stiffness of the bridge and its geometric parameters by FEA. Furthermore, a location-dependent design strategy for the stretchability of bridges is given by geometric mapping.

2. Conformal Criterion for Island

2.1. Conformal Modelling for Island

An island-bridge structure array is mapped onto a hard, non-developable surface, as shown in Figure 1a. The islands in the array are quite small compared to the target surface, so it is reasonable to use a small surface to approximate the local target surface covered by the island. Here, a torus surface under control by two principal curvatures, κ_1 and κ_2, is chosen for theoretical study. $\alpha = \kappa_1/\kappa_2$ is a geometric parameter that controls the shape of the surface. By appointing $|\kappa_1| \leq |\kappa_2|$, α is fixed among -1 and 1, which simplifies the analysis greatly. Different kinds of surfaces can be described by tuning α, such as saddle surfaces (for $-1 \leq \alpha < 0$), cylinders (for $\alpha = 0$), paraboloids (for $0 < \alpha < 1$), and spheres (for $\alpha = 1$). Then, an originally flat elastic island with length of l_{island}, width of w_{island}, and thickness of t_{island} ($t_{island} \ll w_{island} \leq l_{island}$) is mapped onto a part of the torus surface under the assumption that no tension exists in width direction [19], which produces a rectangle conformal zone of length l_{island}. Let the coordinates x and y denote the distance from the island center along the length and width direction, respectively, as shown in Figure 1b. Figure A1 shows the situation when the width direction is deviated from the bending direction of curvature κ_2 with a deflection angle θ. The relationship between conformal strain energy on the island and deflection angle θ is shown in Figure A2. It can be seen from the result that the island has lowest strain energy when $\theta = 0$, which means a most steady state. So, we adopt this state to perform the analysis.

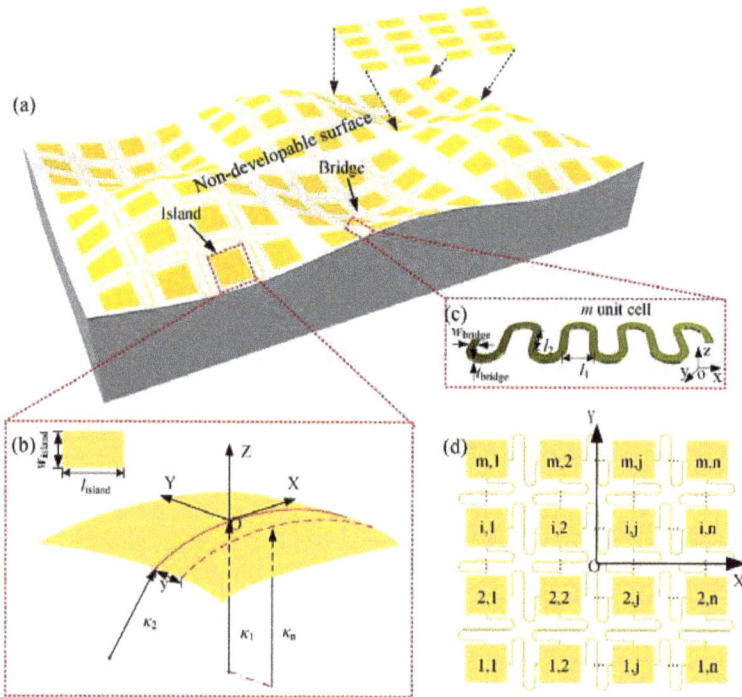

Figure 1. (**a**) An island-bridge structure array on a non-developable surface; (**b**) theory model of island on a torus surface under control by two principal curvatures, κ_1 and κ_2; (**c**) schematic of geometric parameters for a serpentine bridge with *m* unit cells; (**d**) a numbered island-bridge structure array with *m* rows and *n* columns of islands.

The strain in island for above problem is given as follows [36]:

$$\varepsilon_{xx} = \frac{\alpha \kappa_2 z \cos(\kappa_2 y)}{1+[\cos(\kappa_2 y)-1]\alpha} + [\cos(\kappa_2 y) - 1]\alpha$$
$$\varepsilon_{yy} = \kappa_2 z \tag{1}$$
$$\varepsilon_{xy} = 0$$

For the conciseness of energy integration, Equation (1) is replaced by an approximate one Equation (2) by Taylor expanding the $\cos(\kappa_2 y)$. The error between Equation (1) and Equation (2) is below 2% when the non-dimensional width $\kappa_2 w_{island} \leq 0.5$, as shown in Figure A3a, which is reasonable for engineering application.

$$\varepsilon_{xx} = -\tfrac{1}{2}\alpha\kappa_2^2 y^2 + \alpha\kappa_2 z$$
$$\varepsilon_{yy} = \kappa_2 z \tag{2}$$
$$\varepsilon_{xy} = 0$$

When $\alpha < 0$, the surfaces become saddles, and stretching strain will be produced in the island. With the increase of $\kappa_2 w_{island}$, the maximum strain in the island may exceed the failure strain 1% for most of inorganic materials on electronic applications, such as silicon [37] and zinc oxide [38], as shown in Figure 2a. When $\alpha > 0$, paraboloids are described; then, compressing strain shows up and may cause island buckling or failure. When the value of α decreases from 1 to 0, the surface tends to be a cylinder, and its developability is promoted, so the geometric mismatch strain in island is reduced. Specifically, when $\alpha = 0$, the surfaces will turn into a cylinder upon which the island is under pure bending in y direction.

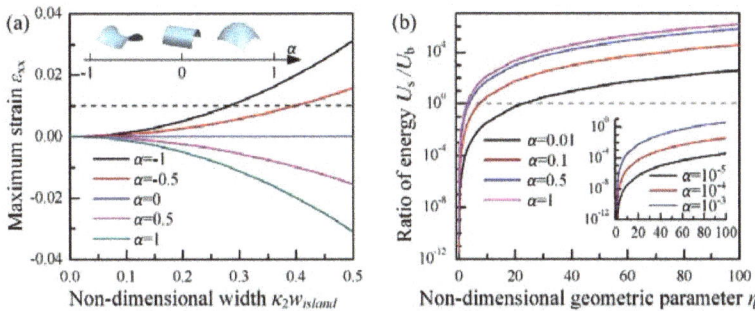

Figure 2. Strain and strain energy in the island during conformal contact: (**a**) maximum strain in island with non-dimensional width $\kappa_2 w_{island}$ at $\kappa_2 t_{island} = 10^{-6}$; (**b**) the ratio of stretching strain energy to bending strain energy with non-dimensional parameter η.

By integrating Equation (2) in island domain, the strain energy on island is given below

$$
\begin{aligned}
U_{strain} &= \tfrac{1}{2}\bar{E}_{island} l_{island} \int_{-w_{island}/2}^{w_{island}/2} \int_{-t_{island}/2}^{t_{island}/2} \left[\varepsilon_{xx}^2(y,z) + \varepsilon_{yy}^2(y,z) + 2\nu\varepsilon_{xx}(t,z)\varepsilon_{yy}(y,z)\right] dy dz \\
&= \frac{E_{island} l_{island} w_{island} t_{island}^3 \kappa_2^2 (1 + \alpha^2 + 2\nu\alpha)}{24(1 - v_{island}^2)} + \frac{E_{island} l_{island} t_{island} w_{island}^5 \kappa_2^4 \alpha^2}{640(1 - v_{island}^2)}
\end{aligned}
\tag{3}
$$

where $\bar{E}_{island} = E_{island}/(1 - v_{island}^2)$, E_{island}, and v_{island} are the Young modulus and Poisson's ratio of the island, respectively.

Figure A3b shows an error of less than 2% when $\kappa_2 w_{island} \leq 0.8$ between the energy solution from Equation (3) and that from numerical integration of strain in Equation (1), which is acceptable for engineering application. The first item in Equation (3) is the energy contribution from bending (indicated as U_b) and the second one is from stretching (indicated as U_s). The ratio of energy contribution of those two deformations produces a non-dimensional geometric parameter $\eta = w_{island}\sqrt{\kappa_2/t_{island}}$. When η and α is small, the surface is nearly developable, so the bending energy is primary. With the increase of η, stretching behavior will contribute more to conformal energy and become dominant eventually, as shown in Figure 2b.

According to the principle of energy minimization, conformal contact is stable when $dU_{strain}/dw_{island} \leq \gamma l_{island}$, which implies

$$|\kappa_2|w_{critical1} = \sqrt[4]{\frac{128(1 - v_{island}^2)}{\alpha^2}} \frac{\gamma}{E_{island}t_{island}} - \frac{16(|\kappa_2|t_{island})^2(1 + \alpha^2 + 2v_{island}\alpha)}{3\alpha^2} \tag{4}$$

where $|\kappa_2|w_{critical1}$ is the non-dimensional maximum critical conformal width from energy minimization and $\gamma/E_{island}t_{island}$ is the non-dimensional interface adhesion energy per unit area.

Meanwhile, the maximum strain in the island should not exceed the failure strain of functional materials on it, so that the electronics can keep working after being transferred to non-developable surface, which implies

$$\varepsilon_{max} = |\alpha||\kappa_2|\frac{t_{island}}{2} + \frac{1}{8}|\alpha|\kappa_2^2 w_{island}^2 \leq \varepsilon_{critical} \tag{5}$$

where $\varepsilon_{critical}$ is the critical failure strain of functional material on island.

Hence, the maximum critical conformal width given by material limit is

$$|\kappa_2|w_{critical2} = \sqrt{\frac{8\varepsilon_{critical} - 4|\alpha||\kappa_2|t_{island}}{|\alpha|}} \tag{6}$$

By comparing two critical widths mentioned above, the final critical width for conformal is given by

$$|\kappa_2|w_{critical} = min(|\kappa_2|w_{critical1}, |\kappa_2|w_{critical2}) \tag{7}$$

Referring to the curvilinear electronics system applications, they usually have mm-wide and μm-thick islands; a big enough η is almost satisfied, which implies that the stretching energy is dominant in conformal strain energy, so Equation (7) can be rewritten with

$$
\begin{aligned}
|\kappa_2|w_{critical} &= \sqrt[4]{\frac{128\left(1-v_{island}^2\right)}{\alpha^2}\frac{\gamma}{E_{island}t_{island}}} \quad &when\ \xi \leq \xi_{critical} \\
|\kappa_2|w_{critical} &= \sqrt{\frac{8\varepsilon_{critical}}{|\alpha|}} \quad &when\ \xi > \xi_{critical}
\end{aligned}
\tag{8}
$$

where $\xi = \gamma/\left(E_{island}t_{island}\varepsilon_{critical}^2\right)$ is a non-dimensional parameter to comprehensively evaluate the effects of the adhesion energy and the failure strain of the material, and $\xi_{critical} = 1/\left[2\left(1-v_{island}^2\right)\right]$ is constant given by making the above two critical widths equal (it depended only on the Poisson's ratio of the island).

The line $\xi = \xi_{critical}$ divides the conformal domain into two regions, so-called 'weak adhesion' and 'strong adhesion', as shown in Figure 3. In the weak adhesion region ($\xi < \xi_{critical}$), the critical conformal width increases with adhesion at the interface, which is consistent with the result given by Majidi [12]. Once $\xi > \xi_{critical}$, it moves into the strong adhesion region. In this region, the critical conformal width is decided by material limit and will not increase with adhesion. The failure mechanisms of island in those two regions are quite different. In the week adhesion region, the maximum strain on the island remains below the failure strain during conformal contact, and detachment will occur at the interface when the adhesion is not able to afford to stable conformal contact. However, in the strong adhesion region, the adhesion is strong enough so that no detachment will happen. With the increase of the width of island, strain in the island will exceed the failure strain and cause the failure of the island eventually.

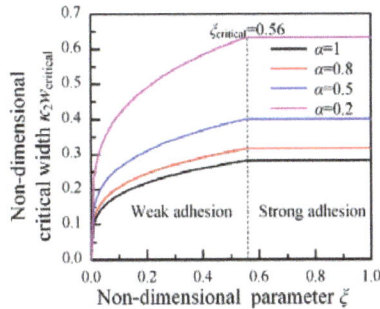

Figure 3. The non-dimensional critical conformal width $\kappa_2 w_{critical}$ with ζ for $\varepsilon_{critical} = 1\%$ and $v_{island} = 0.32$. Two regions, weak adhesion and strong adhesion, are divided by $\zeta_{critical} = 0.56$.

2.2. Adhesion Experiment for Island

Adhesion experiment is performed between polyvinyl chloride (PVC) sticker and Plexiglass sphere with a radius of 50 mm. The PVC sticker is carefully cut into a series of square islands with widths of 10, 15, 20, 25, and 30 mm ($\kappa_2 w_{island} = 0.2, 0.3, 0.4, 0.5, 0.6$) by a cutting machine. Prior to the experiment, the spherical surfaces are scrubbed with alcohol and then air dried. Next, the PVC square island is slowly peeled off from the release substrate and pre-attached to the Plexiglass sphere to make sure that the center of the island is aligned with the sphere center. A soft stamp is used to apply pressure on the top of the PVC island to help further conformal contact. Here, a sponge is used as a stamp due to its negligible traction to the island. Finally, the stamp is removed slowly from the sphere, and the critical conformal width is measured after the conformal region remaining stable. When $\kappa_2 w_{island} < 0.3$, the island completely conforms to the sphere, and no detachment is observed, as shown in Figure 4a,b. As $\kappa_2 w_{island}$ gets bigger, detachments will show up on both sides of the island (Figure 4c), and then the four sides of the island (Figure 4d,e). Due to the compressive strain in the island, the detached parts turn into buckling waves, as shown in Figure 4f. It is interesting to find that the bigger the width of the island is, the more buckling waves will be produced.

The thickness of the PVC sticker is measured by laser scanning confocal microscope (VK-X200, KEYENCE, Osaka, Japan), and a total thickness of 100 µm is given. Then, tension tests and peel tests are performed by a universal mechanical tester (INSTRON 5944, Instron, Norwood, MA, USA) and home-made peel platform to give Young's modulus E_{island}, Poisson's ratio v_{island}, yield strain, and work of adhesion γ. The operational processes and test results for tension test and peel test are listed in Appendix C. For the adhesive PVC sticker used in the experiment, Young's modulus, Poisson's ratio, yield strain, and work of adhesion γ are $E_{island} = 1.29$ GPa, $v_{island} = 0.32$, $\varepsilon_{critical} = 2\%$, and $\gamma = 7.596$ N/m, respectively. The non-dimensional parameter ζ for this experiment is 0.147, which is less than the critical one, which corresponds to a weak adhesion condition. The theoretical non-dimensional critical conformal width given by the first equation in Equation (8) is 0.2868, which is quite close to the experimental one ($\kappa_2 w_{critical} = 0.3$). It is worth noting that with the further increase of the width of island after $\kappa_2 w_{island} > 0.4$, the width of the conformal region will reduce (for $\kappa_2 w_{island} = 0.5$, the conformal width is 0.26 and for $\kappa_2 w_{island} = 0.6$, the conformal width is 0.24), which may come from the influence of the un-conformal region. As the width of island gets bigger, the un-conformal region gets bigger too, so the strain energy in the un-conformal region will be larger and larger. However, for the problem solving the critical width, this part of energy is not under consideration.

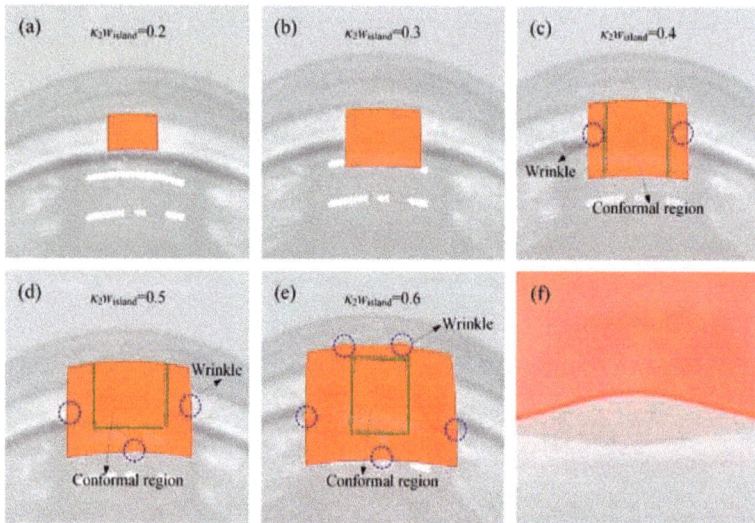

Figure 4. The conformal behaviors between sphere and PVC islands with different width: (a) $\kappa_2 w_{critical} = 0.2$, (b) $\kappa_2 w_{critical} = 0.3$, (c) $\kappa_2 w_{critical} = 0.4$, (d) $\kappa_2 w_{critical} = 0.5$, (e) $\kappa_2 w_{critical} = 0.6$, and (f) enlarge view of wrinkle in (c).

3. Mechanics of Stretchable Bridges

The design demands for bridges include two aspects. First of all, the tensile stiffness of the bridge needs to be designed so that the island can be deterministically assembled onto the target surface. On the other hand, the stretchability of the bridge needs to be designed to bear strain produced during the conformal process. In this section, the dependency of tensile stiffness of bridge on its geometric parameters and the demand for stretchability of the bridge on sphere are studied by FEA simulation and geometric mapping, respectively.

3.1. Tensile Stiffness Design for Bridges

A serpentine bridge with m unit cells is taken into consideration, as shown in Figure 1c. Each unit cell is composed of two half circles and two straight lines with length l_2 and spacing l_1 and has a rectangular cross section with width w_{bridge} and thickness t_{bridge}. The serpentine bridge made of single layer PI with Young's modulus $E_{PI} = 2.5$ GPa and Poisson's ratio $\nu_{PI} = 0.34$ is analyzed to given the scaling laws of axis force, and its dependency on the geometric parameters mentioned above. The tensile stiffness can be solved by taking a derivative of the axis force with respect to axial displacement. The serpentine bridge is clamped at two ends and pulls from an axial direction (x direction in Figure 1c). Four-node shell elements are used to model the serpentine bridge, and high-quality meshes are adopted to guarantee the accuracy of those analyses. A two-step method is used for FEA simulations. Firstly, the buckling analysis is adopted to get buckling strain and buckling modes for the serpentine bridge. Then, using the buckling modes from step 1 as initial imperfection to continue a nonlinear static analysis, a small enough damping is added to the model to ensure the convergence of the analysis.

The relationship between the axial force and apply strain reveals typical 'J-shape' stress-strain behavior, as shown in Figure 5a. The deformation of the serpentine bridge with strain shows a three stage, and two transition point are observed in simulations. The first stage is when the apply strain is lower than the critical strain for buckling. In this stage, only in-plane deformations exist, as shown in Figure 6a. The fact that the axial force keeps a linear relation with strain implies a constant tensile stiffness in this stage. The second stage starts with the buckling of the bridge when $\varepsilon_{appl} \geq 22\%$, as illustrated in Figure 6b. In this stage, the bridge undergoes complicated bending and twisting deformation, and by comparing the configuration of the bridge in Figure 6c,d, it can be found that the tension between two ends is mainly matched by the rotation of the straight lines in the bridge. Hence, the tensile stiffness is in decline and maintains a constant approximately. With the increase of strain, it enters into the third stage. In this stage, the arc in the bridge begins to be straightened as shown in Figure 6e,f, so the tensile stiffness increases sharply with strain.

Figure 5b shows the relationship between the axial force and the number of unit cells *m*. With the increase of *m*, axial force at the end of the bridge is decreased, which means a smaller tensile stiffness as well. Additionally, the effects of *m* tends to be saturated at *m* = 6. Figure 5c show a very good linear correlation between the axial force and the third power of the thickness of the bridge, and the same law can be seen in Figure 5d with the third power of the width, which corresponds to the contributions of out-plane and in-plane deformation, respectively.

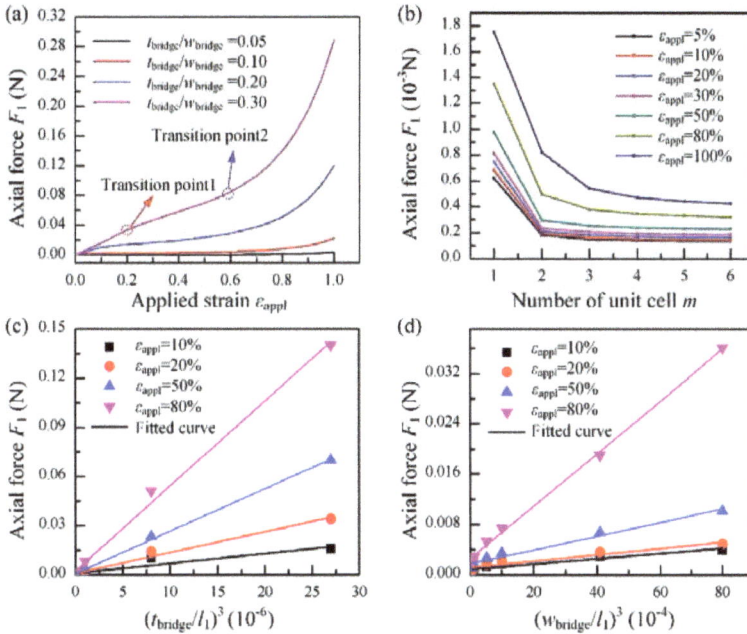

Figure 5. The axial force of a serpentine interconnect under stretching, obtained from the finite element analysis with different parameters: (**a**) applying strain, (**b**) wave numbers of bridge, (**c**) thickness of bridge, (**d**) width of bridge.

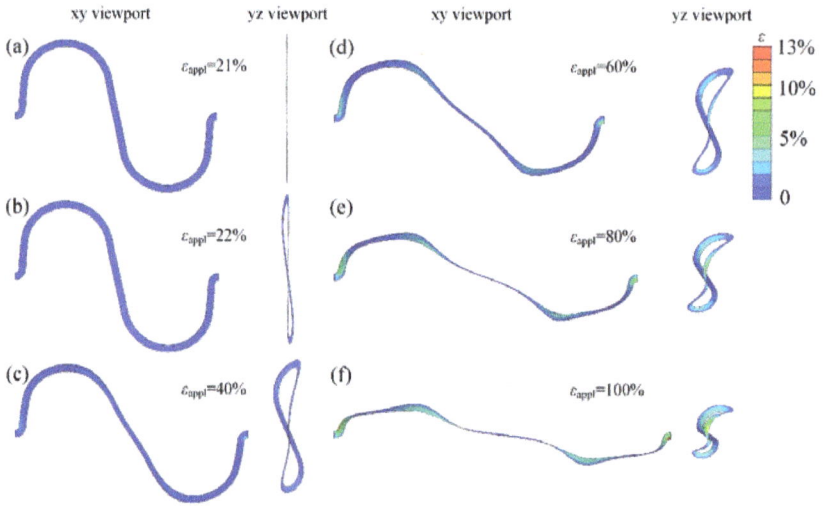

Figure 6. The max principal strain in the bridge versus the applied strain, and the corresponding deformation configurations in xy and yz viewport: (**a**) ε_{appl} = 21%, (**b**) ε_{appl} = 22%, (**c**) ε_{appl} = 40%, (**d**) ε_{appl} = 60%, (**e**) ε_{appl} = 80%, and (**f**) ε_{appl} = 100%.

3.2. Stretchability Demands for Bridges

In this section, an $m \times n$ array of island-bridge structure with single island size w_{island} and distance s between two islands is mapped onto a sphere with radius R. The total length and width of the array are $l_{tot} = n\, w_{island} + (n-1)s$ and $w_{tot} = m\, w_{island} + (m-1)s$, respectively.

For the island numbered as (i, j), its center coordinates are

$$X_{i,j} = \frac{2j-1}{2} w_{island} + (j-1)s - \frac{l_{tot}}{2}$$
$$Y_{i,j} = \frac{2i-1}{2} w_{island} + (i-1)s - \frac{w_{tot}}{2} \tag{9}$$

Considering the unstretchable nature of the island, the demands for stretchablity of the vertical bridges during conformal processing can be given as follows by comparing the coordinates before and after the mapping

$$\varepsilon_{i,j \to i+1,j} = \frac{1}{6} \frac{w_{island} + s}{s} \frac{\left(Y^2_{i+1,j} + Y_{i+1,j}Y_{i,j} + Y^2_{i,j}\right)}{R^2 - X^2_{i,j}} \tag{10}$$

where $i,j \to i+1,j$ means the bridge that connects two islands numbered as (i, j) and $(i + 1, j)$ respectively.

In a similar way, the demands for stretchablity of the horizontal bridges are

$$\varepsilon_{i,j \to i,j+1} = \frac{1}{6} \frac{w_{island} + s}{s} \frac{\left(X^2_{i+1,j} + X_{i+1,j}X_{i,j} + X^2_{i,j}\right)}{R^2 - Y^2_{i,j}} \tag{11}$$

Two parameters, functional coverage $\eta_1 = w_{island}/(w_{island} + s)$ and area coverage $\eta_2 = w_{tot}l_{tot}/\pi R$, are defined to describe the area ratio of sensor elements to the whole device and the device to the target surface, respectively. Figure 7 shows location-dependent demands for stretchability of horizontal bridges in the array. The same law is existent for vertical bridges as well. It is found that the bridges far away from the center of the device have higher demands for stretchability than those nearby. Hence, there are two design strategies for stretchability of bridge: one is using the maximum stretchability demand for all bridges in the array and another one is to design different stretchabilities for bridges at different locations. The former may be a convenient way, but as shown in Figure 7b, the demands for stretchability increase with η_1 sharply, and for $\eta_1 = 0.8$, there is 14% difference in numerical value between bridges at the edge and those near center, so this strategy will produce much redundancy in the whole device. However, a narrower wire width is usually needed for higher stretchability, which means a higher resistance as well. Hence, the latter strategy may be a more economical way.

Figure 7c shows that the maximum stretchability demand increases with the number of islands in the array at the same functional coverage η_1 and area coverage η_2, and it tends to converge to a constant finally. For an array with larger η_1, such as $\eta_1 = 0.8$, the effect of the number of islands is more obvious, which implies that when we try to gain better conformability of the device by reducing the size of island, a higher stretchability will be needed. On the other hand, if the device is needed to cover a larger target surface with high functional coverage to obtain better performance, a higher stretchability will be needed, as shown in Figure 7d.

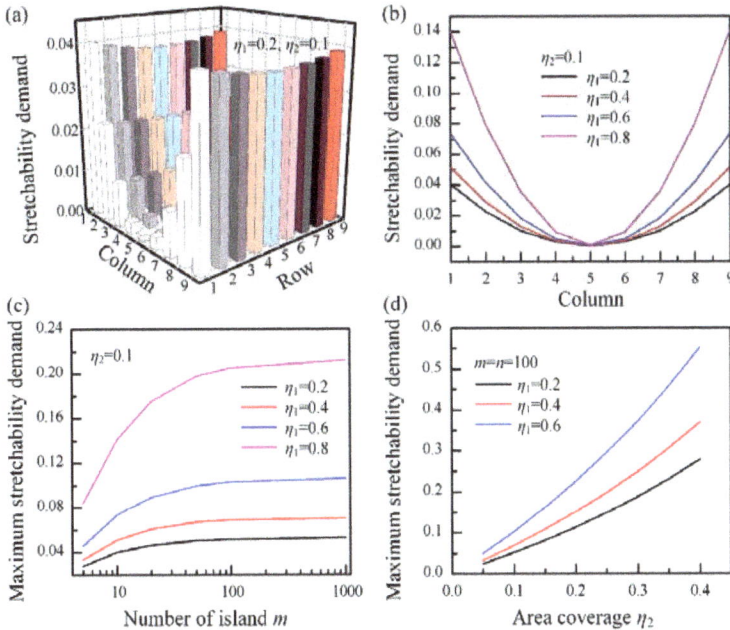

Figure 7. Demands for stretchability of the bridges given by geometric method: location-dependent property of demands for stretchability in the array (**a**) and at the first row (**b**) for the horizontal bridges; maximum demand for stretchability of the device with the number of islands (**c**) and area coverage (**d**).

4. Conclusions

In this work, a theoretical model for the island conformed to a torus surface, governed by two principal curvatures κ_1 and κ_2 was set up. By adjusting the ratio of two principal curvatures, denoted as α, the conformal problem for island on saddle surface, cylinder, paraboloid, and sphere can be described. A non-dimensional critical conformal width was given for the island as a function of non-dimensional interfacial adhesion energy per unit area $\gamma/E_{island}t_{island}$ and non-dimensional thickness for the island $\kappa_2 t_{island}$ and α by combining the principle of energy minimization and the limit of material failure. A Poisson's ratio relevant critical value $\zeta_{critical}$ divides the conformal domain into two regions, in which the adhesion and the limit of material failure are in charge, respectively. Besides, the relationships between the axial force of the bridge and its geometric parameters were revealed by FEA method so that the tensile stiffness of the bridge could be predesigned to help guide the deterministic assembly. Finally, a location-dependent demand for the stretchability of the bridge was found, and, based on this, an economical strategy was proposed by designing different stretchabilities for the bridge according to its location. Higher stretchability is a guarantee of better conformability. However, there are contradictions between stretchability and electrical performance; the collaborative optimization design is yet to be studied.

Author Contributions: L.X. performed the theoretical derivation and wrote the manuscript. Y.A.H. shared the research ideas and methods and wrote this paper. All authors discussed the results and conclusions on the manuscript. All authors read and approved the final manuscript.

Funding: This research was funded by the National Natural Science Foundation of China (51635007), Special Project of Technology Innovation of Hubei Province (2017AAA002), and Program for HUST Academic Frontier Youth Team.

Acknowledgments: The authors would like to thank Flexible Electronics Manufacturing Laboratory in Comprehensive Experiment Center for Advanced Manufacturing and Equipment Technology.

Conflicts of Interest: The authors declare no conflict of interest.

Appendix A. Conformal Strain Energy with Angle of Deviation

Let coordinates OXYZ and OX'Y'Z, referring to island and surface, respectively, and the angle between them is marked as θ, as shown in Figure A1.

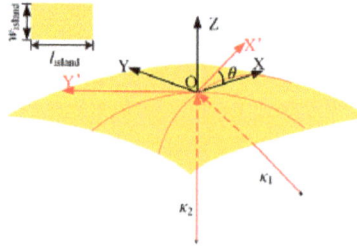

Figure A1. The schematic for island on surface with initial angle θ.

The strain components in coordinate OX'Y'Z are given by Equation (A1) as below

$$\varepsilon_{x'x'} = \frac{\alpha\kappa_2 z \cos(\kappa_2 y')}{1+[\cos(\kappa_2 y')-1]\alpha} + [\cos(\kappa_2 y') - 1]\alpha$$
$$\varepsilon_{y'y'} = \kappa_2 z \tag{A1}$$
$$\varepsilon_{x'y'} = 0$$

where $\begin{bmatrix} x' \\ y' \\ z' \end{bmatrix} = T \begin{bmatrix} x \\ y \\ z \end{bmatrix}, T = \begin{bmatrix} c & -s & 0 \\ s & c & 0 \\ 0 & 0 & 1 \end{bmatrix}$ is the coordinate transformation matrix from coordinate OXYZ to coordinate OX'Y'Z, $c = \cos\theta$ and $s = \sin\theta$.

The strain components in coordinate OXYZ can be solved from Equation (A1) by coordinate transformation

$$
\begin{bmatrix} \varepsilon_{xx} \\ \varepsilon_{yy} \\ \varepsilon_{xy} \end{bmatrix} = A^{-1} \begin{bmatrix} \varepsilon_{x'x'} \\ \varepsilon_{y'y'} \\ \varepsilon_{x'y'} \end{bmatrix} \tag{A2}
$$

where $A = \begin{bmatrix} c^2 & s^2 & 2sc \\ s^2 & c^2 & -2sc \\ -sc & sc & c^2 - s^2 \end{bmatrix}$ is the strain transformation matrix from coordinate OXYZ to

coordinate OX'Y'Z.

Substituting Equation (A2) into Equation (3) and integrating the island domain using the numerical method, the strain energy on the island is given.

Figure A2 shows the relationship between conformal strain energy per unit area in the island and initial angle θ. When $w_{island}/l_{island} < 1$, the minimum energy shows up when $\theta = 0$, which means that when the width direction is aligned with the bending direction of maximum curvature κ_2, the conformal status is most stable. This is why we have made this appointment in Section 2. When $w_{island}/l_{island} = 1$, the minimum energy shows up in both $\theta = 0°$ and $\theta = 90°$; this is reasonable because of the symmetry of the square island.

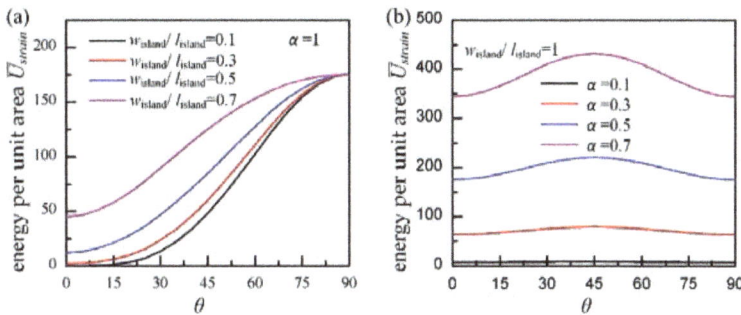

Figure A2. The conformal strain energy per unit area in island with initial angle θ at different width-length ratios (**a**) and different curvature ratios (**b**).

Appendix B. Relative Error between Approximate Solution and Exact Solution

To simplify the integral in Equation (3) and give a concise solution to critical conformal width, the strain in Equation (1) is approximated with Equation (2) using the Taylor expansion of $\cos(\kappa_2 y) = 1 - \frac{1}{2}(\kappa_2 y)^2 + o\left((\kappa_2 y)^4\right)$ when $|\kappa_2 y| < 1$. Figure A3 shows the relative error between approximate solution Equation (1) and accurate solution Equation (2). The accurate solution for strain energy is obtained by numerical integration from Equation (1). When $\alpha = 0$, the surface turns into a cylinder, and there is no error in the approximate solution. When $\alpha \neq 0$, the relative error increases with non-dimensional width $\kappa_2 w_{island}$. Here, we choose 2% as the up limit of the error; thus, $\kappa_2 w_{island} \leq 0.5$.

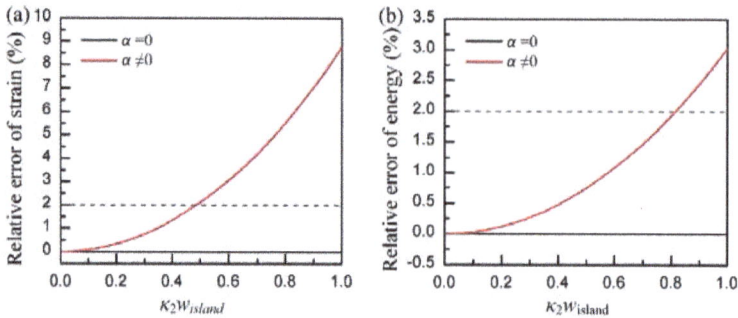

Figure A3. The relative error between approximate solution and accurate solution with $\kappa_2 w_{island}$: (**a**) for strain and (**b**) for conformal strain energy.

Appendix C. Material Parameter Test

The tension test is performed by a universal mechanical tester (INSTRON 5944, Instron, Norwood, MA, USA), as shown in Figure A4a. A dumbbell-shaped sample of PVC sticker with a length of 80 mm and a width of 5 mm in test section is connected to the tester by two pneumatic clampings. Then, the Z translation stage is moved upwards slightly to tighten the sample before the test. Finally, resetting the value of the load cell to zero and starting test with a tension rate of 1 mm/s. Figure A4c shows the stress-strain curve of the PVC sticker, and the slope of the curve, that is Young's modulus, given by a linear fit of the elastic segment is 1.296 GPa. Another two samples are tested using the same method, and the Young's modulae given by those tests are 1.298 GPa and 1.276 GPa, respectively. A mean value of 1.290 GPa is given as the Young's modulus for PVC sticker. Figure A4c also shows a yield strain of approximate 2% for PVC sticker. Besides, the Poisson's ratio is estimated from the percentage of cross-section contraction to its elongation.

Figure A4b shows a home-made peel platform whose peel angle and peel rate can be controlled. More detail about this platform can be found in [39]. A rectangle sample of PVC sticker with a length of 200 mm and a width of $w_{PVC} = 20$ mm is prepared for test. Firstly, the sample is adhered to a Plexiglass plate, which is cleaned by alcohol and fixed at an angle-adjustable jig. Secondly, the sample is adjusted to the location right under the load cell by X and Y-motion, and the peel angle β is fixed at 135°. Thirdly, a part of the sample is peeled and its end is connected to the fixture of the load cell. After that, the location of the sample is adjusted again by X and Z-motion to make sure that the detached part of the sample remains upright. Finally, the value of the load cell is reset to zero and test is started with a peel rate of 1 mm/s. The force in the load cell starts at zero and keeps increasing before the beginning of the peeling at the interfaces, as shown in Figure A4d. The oscillation in force may come from the heterogeneous adhesion in PVC sticker or the motion of linear stepping motor; thus, the mean value of force in the peel segment is used as peel force. Three samples are tested to give an average peel force $F_{peel} = 0.089$ N. Referring to the equation given in [39], the work of adhesion per unit area is $\gamma = F_{peel}(1 - \cos \beta)/w_{PVC} = 7.596$ N/m.

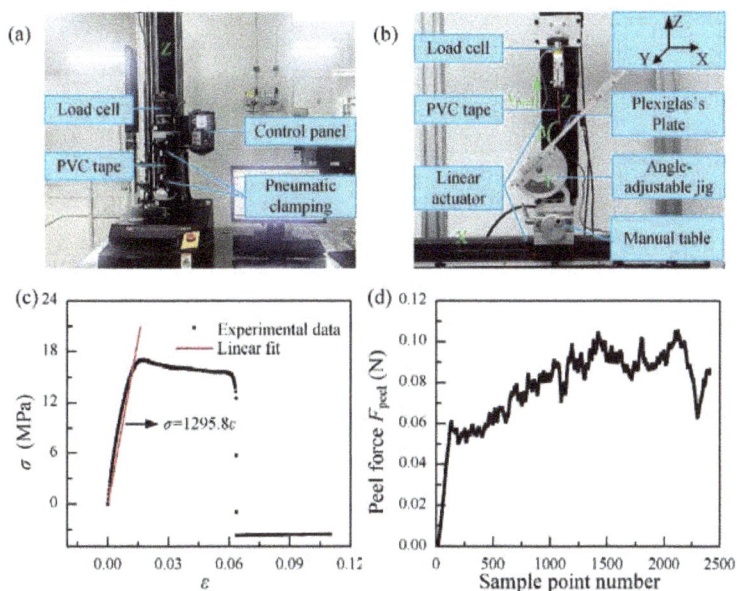

Figure A4. Experimental apparatus and experimental data for material parameter test: (**a**) an universal mechanical tester (INSTRON 5944); (**b**) a home-made peel platform with an angle-adjustable jig, in which the X or Z-motion of the translation stage is able to be driven by two independent linear/electric actuators, and its Y-motion depends on a manually single axis table; (**c**) stress-strain curve for PVC sticker in tension test; and (**d**) peel force for PVC sticker in peel test with peel angle $\beta = 135°$ and peel rate $v_{peel} = 1$ mm/s.

References

1. Jung, I.; Xiao, J.; Malyarchuk, V.; Lu, C.; Li, M.; Liu, Z.; Yoon, J.; Huang, Y.; Rogers, J.A. Dynamically tunable hemispherical electronic eye camera system with adjustable zoom capability. *Proc. Natl. Acad. Sci. USA* **2011**, *108*, 1788–1793. [CrossRef] [PubMed]

2. Ko, H.C.; Stoykovich, M.P.; Song, J.; Malyarchuk, V.; Choi, W.M.; Yu, C.-J.; Geddes, J.B.; Xiao, J.; Wang, S.; Huang, Y.; et al. A hemispherical electronic eye camera based on compressible silicon optoelectronics. *Nature* **2008**, *454*, 748–753. [CrossRef] [PubMed]

3. Song, Y.M.; Xie, Y.; Malyarchuk, V.; Xiao, J.; Jung, I.; Choi, K.-J.; Liu, Z.; Park, H.; Lu, C.; Kim, R.-H.; et al. Digital cameras with designs inspired by the arthropod eye. *Nature* **2013**, *497*, 95–99. [CrossRef] [PubMed]

4. Xu, L.; Gutbrod, S.R.; Bonifas, A.P.; Su, Y.; Sulkin, M.S.; Lu, N.; Chung, H.-J.; Jang, K.-I.; Liu, Z.; Ying, M.; et al. 3D multifunctional integumentary membranes for spatiotemporal cardiac measurements and stimulation across the entire epicardium. *Nat. Commun.* **2014**, *5*, 3329. [CrossRef] [PubMed]

5. Xu, L.; Gutbrod, S.R.; Ma, Y.; Petrossians, A.; Liu, Y.; Webb, R.C.; Fan, J.A.; Yang, Z.; Xu, R.; Whalen, J.J.; et al. Materials and fractal designs for 3D multifunctional integumentary membranes with capabilities in cardiac electrotherapy. *Adv. Mater.* **2015**, *27*, 1731–1737. [CrossRef] [PubMed]

6. Son, D.; Lee, J.; Qiao, S.; Ghaffari, R.; Kim, J.; Lee, J.E.; Song, C.; Kim, S.J.; Lee, D.J.; Jun, S.W.; et al. Multifunctional wearable devices for diagnosis and therapy of movement disorders. *Nat. Nanotechnol.* **2014**, *9*, 397–404. [CrossRef] [PubMed]

7. Someya, T.; Kato, Y.; Sekitani, T.; Iba, S.; Noguchi, Y.; Murase, Y.; Kawaguchi, H.; Sakurai, T. Conformable, flexible, large-area networks of pressure and thermal sensors with organic transistor active matrixes. *Proc. Natl. Acad. Sci. USA* **2005**, *102*, 12321–12325. [CrossRef] [PubMed]

8. Dagdeviren, C.; Su, Y.; Joe, P.; Yona, R.; Liu, Y.; Kim, Y.S.; Huang, Y.; Damadoran, A.R.; Xia, J.; Martin, L.W.; et al. Conformable amplified lead zirconate titanate sensors with enhanced piezoelectric response for cutaneous pressure monitoring. *Nat. Commun.* **2014**, *5*, 4496. [CrossRef] [PubMed]
9. Li, Y.; Samad, Y.A.; Liao, K. From cotton to wearable pressure sensor. *J. Mater. Chem. A* **2015**, *3*, 2181–2187. [CrossRef]
10. Samad, Y.A.; Li, Y.; Alhassan, S.M.; Liao, K. Novel graphene foam composite with adjustable sensitivity for sensor applications. *ACS Appl. Mater. Interfaces* **2015**, *7*, 9195–9202. [CrossRef] [PubMed]
11. Boland, C.S.; Khan, U.; Ryan, G.; Barwich, S.; Charifou, R.; Harvey, A.; Backes, C.; Li, Z.; Ferreira, M.S.; Mobius, M.E.; et al. Sensitive electromechanical sensors using viscoelastic graphene-polymer nanocomposites. *Science* **2016**, *354*, 1257–1260. [CrossRef] [PubMed]
12. Lanzara, G.; Salowitz, N.; Guo, Z.; Chang, F.K. A spider-web-like highly expandable sensor network for multifunctional materials. *Adv. Mater.* **2010**, *22*, 4643–4648. [CrossRef] [PubMed]
13. Salowitz, N.; Guo, Z.; Roy, S.; Nardari, R.; Li, Y.-H.; Kim, S.-J.; Kopsaftopoulos, F.; Chang, F.-K. Recent advancements and vision toward stretchable bio-inspired networks for intelligent structures. *Struct. Health Monit.* **2014**, *13*, 609–620. [CrossRef]
14. Kim, D.H.; Lu, N.; Ma, R.; Kim, Y.S.; Kim, R.H.; Wang, S.; Wu, J.; Won, S.M.; Tao, H.; Islam, A.; et al. Epidermal electronics. *Science* **2011**, *333*, 838–843. [CrossRef] [PubMed]
15. Jeong, J.-W.; Kim, M.K.; Cheng, H.; Yeo, W.-H.; Huang, X.; Liu, Y.; Zhang, Y.; Huang, Y.; Rogers, J.A. Capacitive epidermal electronics for electrically safe, long-term electrophysiological measurements. *Adv. Healthc. Mater.* **2014**, *3*, 642–648. [CrossRef] [PubMed]
16. Wang, S.; Li, M.; Wu, J.; Kim, D.-H.; Lu, N.; Su, Y.; Kang, Z.; Huang, Y.; Rogers, J.A. Mechanics of epidermal electronics. *J. Appl. Mech.* **2012**, *79*, 031022. [CrossRef]
17. Dong, W.; Xiao, L.; Zhu, C.; Ye, D.; Wang, S.; Huang, Y.; Yin, Z. Theoretical and experimental study of 2D conformability of stretchable electronics laminated onto skin. *Sci. China Technol. Sci.* **2017**, *60*, 1415–1422. [CrossRef]
18. Li, Y.; Zhang, J.; Xing, Y.; Song, J. Thermomechanical analysis of epidermal electronic devices integrated with human skin. *J. Appl. Mech.* **2017**, *84*, 111004. [CrossRef]
19. Majidi, C.; Fearing, R.S. Adhesion of an elastic plate to a sphere. *Proc. R. Soc. A* **2008**, *464*, 1309–1317. [CrossRef]
20. Hure, J.; Audoly, B. Capillary buckling of a thin film adhering to a sphere. *J. Mech. Phys. Solids* **2013**, *61*, 450–471. [CrossRef]
21. Zhou, Y.; Chen, Y.; Liu, B.; Wang, S.; Yang, Z.; Hu, M. Mechanics of nanoscale wrinkling of graphene on a non-developable surface. *Carbon* **2015**, *84*, 263–271. [CrossRef]
22. Chen, Y.; Ma, Y.; Wang, S.; Zhou, Y.; Liu, H. The morphology of graphene on a non-developable concave substrate. *Appl. Phys. Lett.* **2016**, *108*, 031905. [CrossRef]
23. Hure, J.; Roman, B.; Bico, J. Stamping and wrinkling of elastic plates. *Phys. Rev. Lett.* **2012**, *109*, 054302. [CrossRef] [PubMed]
24. Hure, J.; Roman, B.; Bico, J. Wrapping an adhesive sphere with an elastic sheet. *Phys. Rev. Lett.* **2011**, *106*, 174301. [CrossRef] [PubMed]
25. Mitchell, N.P.; Koning, V.; Vitelli, V.; Irvine, W.T.M. Fracture in sheets draped on curved surfaces. *Nat. Mater.* **2016**, *16*, 89–93. [CrossRef] [PubMed]
26. Kim, D.-H.; Song, J.; Choi, W.M.; Kim, H.-S.; Kim, R.-H.; Liu, Z.; Huang, Y.Y.; Hwang, K.-C.; Zhang, Y.; Rogers, J.A. Materials and noncoplanar mesh designs for integrated circuits with linear elastic responses to extreme mechanical deformations. *Proc. Natl. Acad. Sci. USA* **2008**, *105*, 18675–18680. [CrossRef] [PubMed]
27. Su, Y.; Wang, S.; Huang, Y.; Luan, H.; Dong, W.; Fan, J.A.; Yang, Q.; Rogers, J.A.; Huang, Y. Elasticity of fractal inspired interconnects. *Small* **2015**, *11*, 367–373. [CrossRef] [PubMed]
28. Shi, X.; Xu, R.; Li, Y.; Zhang, Y.; Ren, Z.; Gu, J.; Rogers, J.A.; Huang, Y. Mechanics design for stretchable, high areal coverage gaas solar module on an ultrathin substrate. *J. Appl. Mech.* **2014**, *81*, 124502. [CrossRef]
29. Li, R.; Li, M.; Su, Y.; Song, J.; Ni, X. An analytical mechanics model for the island-bridge structure of stretchable electronics. *Soft Matter* **2013**, *9*, 8476–8482. [CrossRef]
30. Ma, Y.; Feng, X.; Rogers, J.A.; Huang, Y.; Zhang, Y. Design and application of 'j-shaped' stress-strain behavior in stretchable electronics: A review. *Lab Chip* **2017**, *17*, 1689–1704. [CrossRef] [PubMed]

31. Zhang, Y.; Xu, S.; Fu, H.; Lee, J.; Su, J.; Hwang, K.C.; Rogers, J.A.; Huang, Y. Buckling in serpentine microstructures and applications in elastomer-supported ultra-stretchable electronics with high areal coverage. *Soft Matter* **2013**, *9*, 8062–8070. [CrossRef] [PubMed]
32. Dong, W.; Zhu, C.; Ye, D.; Huang, Y. Optimal design of self-similar serpentine interconnects embedded in stretchable electronics. *Appl. Phys. A* **2017**, *123*, 428. [CrossRef]
33. Dong, W.; Wang, Y.; Zhou, Y.; Bai, Y.; Ju, Z.; Guo, J.; Gu, G.; Bai, K.; Ouyang, G.; Chen, S.; et al. Soft human–machine interfaces: Design, sensing and stimulation. *Int. J. Intell. Rob. Appl.* **2018**. [CrossRef]
34. Ma, Q.; Zhang, Y. Mechanics of fractal-inspired horseshoe microstructures for applications in stretchable electronics. *J. Appl. Mech.* **2016**, *83*, 111008. [CrossRef]
35. Zhang, Y.; Huang, Y.; Rogers, J.A. Mechanics of stretchable batteries and supercapacitors. *Curr. Opin. Solid State Mater. Sci.* **2015**, *19*, 190–199. [CrossRef]
36. Chen, Z.; Guo, Q.; Majidi, C.; Chen, W.; Srolovitz, D.J.; Haataja, M.P. Nonlinear geometric effects in mechanical bistable morphing structures. *Phys. Rev. Lett.* **2012**, *109*, 114302. [CrossRef] [PubMed]
37. Kim, D.H.; Xiao, J.; Song, J.; Huang, Y.; Rogers, J.A. Stretchable, curvilinear electronics based on inorganic materials. *Adv. Mater.* **2010**, *22*, 2108–2124. [CrossRef] [PubMed]
38. Yu, K.J.; Yan, Z.; Han, M.; Rogers, J.A. Inorganic semiconducting materials for flexible and stretchable electronics. *npj Flex. Electron.* **2017**, *1*, 4. [CrossRef]
39. Huang, Y.; Liu, H.; Xu, Z.; Chen, J.; Yin, Z. Conformal peeling of device-on-substrate system in flexible electronic assembly. *IEEE Trans. Compon. Packag. Manuf. Technol.* **2018**, *PP*, 1–11. [CrossRef]

micromachines

MDPI

Article

Stretchability—The Metric for Stretchable Electrical Interconnects

Bart Plovie [1,*], Frederick Bossuyt [1,2] and Jan Vanfleteren [1,2,*]

[1]   Department of Electronics and Information Systems, Ghent University, Technologiepark 15,
      9052 Zwijnaarde, Belgium
[2]   IMEC vzw, Kapeldreef 75, 3001 Heverlee, Belgium; frederick.bossuyt@imec.be
[*]   Correspondence: bart.plovie@ugent.be (B.P.); jan.vanfleteren@ugent.be (J.V.); Tel.: +32-9-264-6604 (B.P.)

Received: 2 July 2018; Accepted: 27 July 2018; Published: 1 August 2018

Abstract: Stretchable circuit technology, as the name implies, allows an electronic circuit to adapt to its surroundings by elongating when an external force is applied. Based on this, early authors proposed a straightforward metric: stretchability—the percentage length increase the circuit can survive while remaining functional. However, when comparing technologies, this metric is often unreliable as it is heavily design dependent. This paper aims to demonstrate this shortcoming and proposes a series of alternate methods to evaluate the performance of a stretchable interconnect. These methods consider circuit volume, material usage, and the reliability of the technology. This analysis is then expanded to the direct current (DC) resistance measurement performed on these stretchable interconnects. A simple dead reckoning approach is demonstrated to estimate the magnitude of these measurement errors on the final measurement.

Keywords: stretchability; electronic measurements; stretchable circuits; design metrics; reliability

1. Introduction

Stretchable circuit technology is a recent development, having seen the light of day over the span of the last two decades. It is an attractive solution to common design problems; that is, the need for an electronic circuit to cover large areas or to conform to a moving and deformable device, either during production or in use [1,2].

Before the development of stretchable circuit technology, this was usually solved using long spring-loaded cables, stored in an enclosure that fed and retracted the cable as was necessary to cover the distance [3]. This cable-based approach has the advantage of being incredibly reliable if implemented correctly; additionally, it works from direct current (DC) all the way up to the gigahertz range and offers the capability to handle large currents. However, for all these advantages, this technique has a significant drawback: it requires a large volume for even the smallest number of interconnects. An alternative method, which remains popular and still makes a frequent appearance, is the use of coiled flexible circuit boards to cover these transitions [4]. In fact, both methods are surprisingly common in consumer-applications (e.g., devices with retractable power cords and hard drive reading heads).

In comparison, modern-day stretchable electronics use a more elegant approach to provide high-density interconnects over short to medium distances [5]. These methods are often based on thin, flat substrates embedded in soft elastic polymers, and achieve stretchability through careful structuring or speciality materials [1,6]. These techniques are applied on all levels, ranging from pinhead sized semiconductor devices, all the way to circuit boards covering massive composite structures [1,2,4,7–10].

While some exceptions do exist, most technologies have one feature in common: the lumped circuit elements (e.g., resistors, LEDs, and transistors) remain rigid and non-elastic and are relegated to rigid "islands"; meanwhile, the connections between these islands are made stretchable [7,8,11].

This segregation is possible because most electronic circuits use highly modular designs, with the interconnections between these modules being less critical—meaning that modifying these connections does not affect the circuit's functionality. As a result of this approach, the inherent capabilities of the circuit, regarding stretchability, are quantifiable by investigating the interconnections themselves, or at least that is the assumption commonly made. Arguments exist against this approach, but comparisons including the islands become difficult without considering specific circuit designs. Because of this reason, and this reason alone, only the interconnects themselves are considered in this manuscript, and for large high-density circuits, the effect of the islands should definitely be included.

The traditional approach to compare stretchable circuit technologies is based on considering the maximum percentage increase in length. This metric works by measuring the maximum increase in length the interconnect can survive, and dividing it by its original starting length—measured between the terminals [12]. This simple approach can be a fair indication of technological abilities when considering planar circuits using similar materials and designs, and is generally called "stretchability".

Closely related is the percentage increase in resistance; this is the increase in resistance divided by the resistance measured between the terminals before stretching. It is relatively easy to couple these resistance increases to physical effects, for example, in metal meanders, an increase of resistance tends to indicate the presence of mechanical damage (e.g., necking and micro-cracks) [13]. Whereas in the case of conductive filler particles in a polymer matrix, it can indicate the average proximity of these conductive particles to each other [14]. Upon unloading the stretchable interconnect, which is to say the elongation is reduced to zero, the latter will recover, assuming no defects form in the polymer matrix and no conductive fillers migrate. So, a permanent increase in resistance here would mean the polymer material did not return entirely to its original length, or significant defects formed. The metal meander, on the other hand, will permanently maintain this damage, even though it might not be directly visible because of the metal being pressed together again, which creates an unreliable electrical connection [5,13].

Sadly, both these metrics can introduce a significant amount of bias towards a particular technology, just because of the way these values are measured. What may initially appear to be an insignificant measurement error can quickly add up and lead to incorrect results. To demonstrate the flaws of the above "stretchability" definition, we design, fabricate, and measure a circuit intended to exploit these flaws, and then define a series of alternative metrics to compensate for them. Additionally, the more subtle points of interconnect resistance measurements for stretchable electronics are highlighted.

2. Materials and Methods

The circuit was fabricated using a 246 mm by 207 mm piece of UBE Upisel-N SR-1220 (UBE EXSYMO Co., Ltd., Tokyo, Japan) polyimide flexible copper clad laminate (FCCL) with 18 μm of rolled-annealed copper on 50 μm of polyimide. Using standard printed circuit board processing techniques [4], photolithography and wet-etching, the FCCL was structured into a flexible circuit board (FCB), as illustrated in Figure 1b. This FCB was attached to a 1.6 mm FR-4 carrier board (Hitachi Chemical Company, Ltd., Tokyo, Japan) covered with Taconic FH20LB Tacsil tape (TACONIC, Seongnam-si, Republic of Korea)—a pressure sensitive adhesive for reflow assembly of flexible circuit boards—using a hand roller. The purpose of this adhesive is to form a tacky surface that will hold onto the FCB to prevent entanglement as it is structured into a long meandering (also sometimes referred to as serpentine [15]) interconnect. Once the FCB is attached, the carrier board is placed underneath a 10 kHz-5 W pulsed Nd/YAG laser cutter (OPTEC, Frameries, Belgium). The FCCL is cut at a rate of 5 mm/s without damaging the TacSil tape and carrier board underneath. The material surrounding the desired circuit was peeled away using tweezers and discarded, as shown in Figure 1b, leaving behind the circuit design shown in Figures 1a and 2.

(a)

(b)

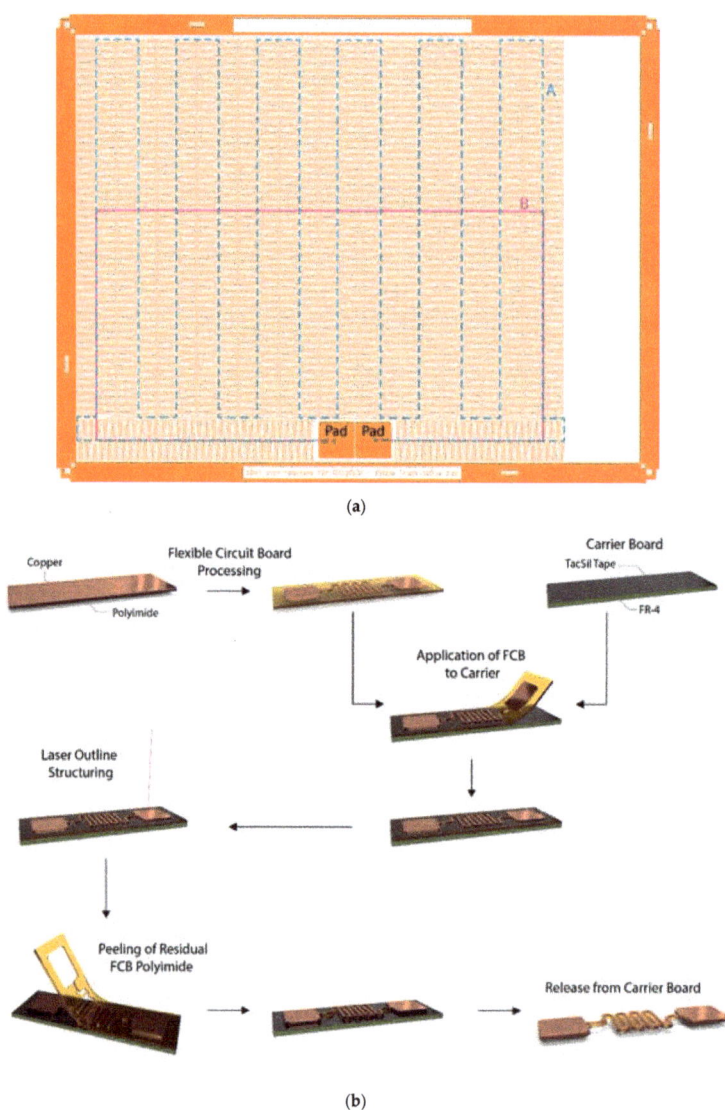

Figure 1. Manufacturing of free-standing stretchable circuits. (**a**) Copper artwork of the manufactured stretchable circuit. (**b**) Process flow used to manufacture the stretchable circuits. First, a flexible copper clad laminate is patterned using photolithography and wet etching. The resulting flexible circuit board (FCB) is then applied to an FR-4 carrier board covered with a pressure-sensitive TacSil Tape adhesive for mechanical support. The outline of the meanders and connection pads is then defined by cutting the flexible circuit board material using a laser cutter. The excess material surrounding the circuit is removed by hand, after which the circuit can be released from the carrier board by carefully peeling it off.

The design of the stretchable interconnect, of which the design file is available as supplementary material, has the goal of optimizing the stretchability. The easiest way to improve the potential increase in length is having this length of conductor available, though this does not necessarily have to be in a practical location. Additionally, placing the pads next to each other increases the achieved stretchability without leading to any practical benefit. A third factor to consider is the reliability, because this is not included in the definition of stretchability, the design can use sharp corners with a small radius, which will hamper reliability. The first step is easily done by placing the 18 mm by 18 mm contact pads against each other. A critical value in this entire operation is the outline spacing, the distance between the edge of the copper artwork and the edge of the polyimide, as indicated in Figure 2. The minimum spacing for the outline of the flexible circuit board in regard to the copper is 100 µm; however, for yield reasons, the outline spacing was increased to 200 µm between the pads, resulting in a distance of 400 µm.

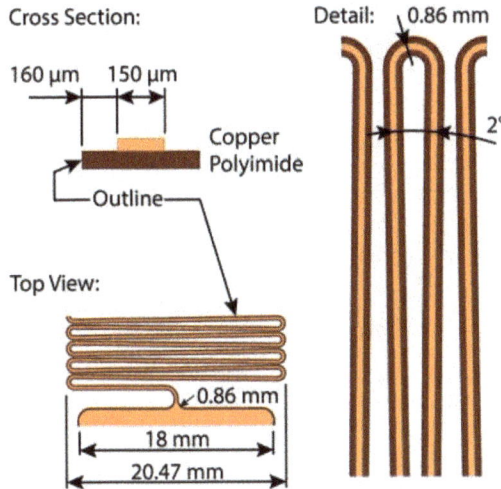

Figure 2. Design details of the stretchable circuit. The side of the polyimide track is the circuit outline. The meanders fold back on themselves at an angle of 2°, optimizing the amount of track in a given surface area without introducing extremely sharp corners.

From these pads, a 150 µm wide copper trace leaves with an outline offset of 160 µm on both sides—as indicated on Figure 2. After the first 90° bend, a U-shaped meander was started, which is folded back on itself—meaning the straight sections connecting the arcs are no longer parallel. This meander has a pitch of 1.5 mm and a height of approximately 20.5 mm and was looped in an accordion pattern above the contact pads. Excluding the length of the contact pads, this meander spans a distance of 66,637.5 mm along its center line. Two additional center lines were defined in Figure 1a, the purpose of which will become clear later on. The first centerline (A), which closely follows the zig-zag pattern, has a length of 2711.7 mm; the second one, however, does not follow this pattern and has a length of only 639.1 mm. The total surface area used by this circuit, including the contact pads, is 50,905.0 mm^2—a number that will become important later on.

Conventional tools, such as a tensile tester, are unable to deal with meanders of these lengths, meaning standard test methods were impossible to use. Instead, the carrier board with the circuit was attached to rigid surface perpendicular to the floor in a long corridor and markings were put on the floor at one-meter intervals using a tape measure, as shown in Figure 3b,c. To determine the resistance of the meander, a Keithley 2400 source meter (Tektronix, Inc., Beaverton, OR, USA) was set up to perform a four-wire measurement with a test current of 1 mA. To extend the test leads, a conventional power cable reel, visible in Figure 3a, was used. To ensure mechanical stability during extension of the meander, one of its pads was attached to the cable reel, as seen in Figure 3a. The reel was then moved backwards at a rate of approximately 10 cm/s to extend the circuit to the desired lengths.

Before measurement, the resistance was allowed to stabilize for at least 30 s to eliminate possible measurement errors due to mechanical vibrations.

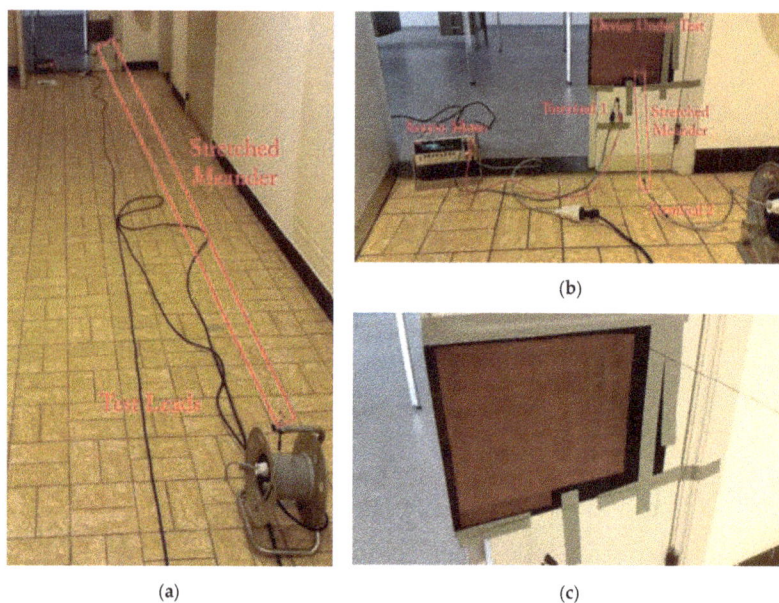

Figure 3. Measurement setup used to determine the meander resistance as a function of the extension: (**a**) measurement setup at the 10 m extension mark; (**b**) measurement setup at the start of the measurement before any extension. The cable reel visible at the bottom right of the image was used to extend the test leads to the desired length; (**c**) detailed view of the carrier board and circuit at the 10 m extension mark.

A second experiment, demonstrating the design dependence of stretchability—even within a given technology—uses a variety of stretchable interconnect designs. For this purpose, 18 designs based on flexible circuit board technology were manufactured and tested until failure using a tensile tester (Instron 5543, Instron, Norwood, MA, USA). As before, the flexible copper clad laminate (Shengyi SF305 101820SR—25 μm of polyimide with 18 μm of adhesively bonded copper) was patterned using industry standard practices. Afterwards, a coverlay film (Shengyi SF305C 1025) was bonded to the resulting flexible circuit board using the manufacturer's recommended press program in a vacuum press (Lauffer RLKV25). The FCB was placed on an identical carrier board and laser structured using a picosecond pulse-length Nd/YAG laser system (3D Micromac microSTRUCT™ vario) to define the meander outline—as illustrated in Figure 1b. The spacing between the copper track, which is 100 μm wide, and the laser path was set to a rather hefty 300 μm—representing the minimum tolerances achievable in an industrial setting. Some of the meanders were encapsulated between two layers of thermoplastic polyurethane (TPU)—100 μm Covestro Platilon U4201 AU—using a vacuum press. To ensure even pressure distribution during the lamination of the TPU, a 1/8-inch silicone–rubber–foam press pad (Rogers Corporation BISCO Foam HT-870) was used between the 25-μm polytetrafluoroethylene (PTFE) release foil and separator plate. The press applies $10 \, \text{N/cm}^2$ at 175 °C for 20 min in a <2 mbar vacuum atmosphere to bond the TPU layers to the polyimide and together. The final samples, shown in Figure 4, were then cut using a guillotine cutter. To enable electrical measurements, the TPU was opened up using a soldering iron, and flexible wires were attached to the samples using leaded solder ($Sn_{63}Pb_{37}$ alloy).

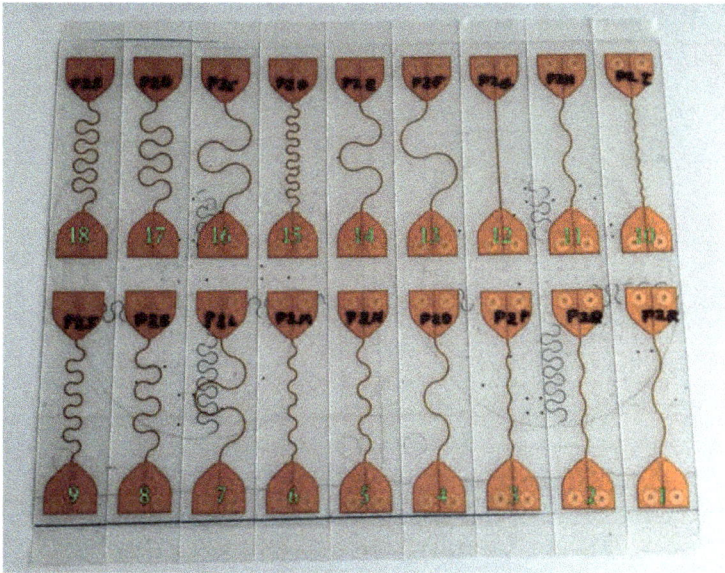

Figure 4. Meander test samples encapsulated in thermoplastic polyurethane using a vacuum press. The black debris visible on the sample originates from FCBs previously cut on the same carrier board.

Each meander has an effective length of 50 mm between the contact pads, of which 3 mm is taken up by the fillets to transition from the connection pads to the meander, as shown in Figure 5. These connections pads at either end of the meander are necessary for both clamping and electrical connectivity—to perform a four-point measurement while the meander is being stretched in a tensile tester. The contact pads were split up and only meet near the start of the meander in an attempt to eliminate their influence on the measurement. The design parameters of these meanders are listed in Table 1, all except design #12 are of the horseshoe-shaped meander type. Design #12 is a reference with a straight track in between the connection pads to determine the inherent capability of the material to meet these deformations. A schematic representation of such a test sample with relevant measurements is visible in Figure 5. Alternatively, the original design file is available as Figure S2.

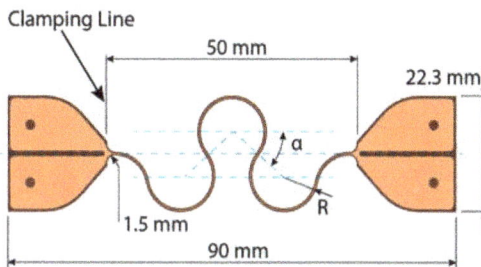

Figure 5. Design of the test vehicle with horseshoe-shaped meanders used for the tensile tests. R is the radius of the horseshoe-shaped meander segment, while α is the opening angle. These two variables combined completely define a horseshoe-shaped meander. To provide a smooth transition to the contact pads, a small 1.5-mm fillet was introduced at both ends. The contact pads at either side of the meander were split up to enable a four-point measurement with separate drive and sense lines.

Table 1. Design parameters for the tested interconnects.

Design	Length [mm]	Radius [mm]	Opening Angle α [°]	Area [mm²]	Segments	Theoretical Stretchability [%]
1	50.605	10.000	−47.330	281.747	3	8
2	49.516	6.547	−56.166	137.118	6	5
3	49.506	4.555	−56.892	102.417	9	5
4	62.037	6.491	−8.238	555.625	3	32
5	62.046	3.649	−13.373	296.565	6	32
6	62.044	2.532	−14.995	209.313	9	32
7	97.147	6.678	39.037	1056.007	3	107
8	97.063	3.729	29.092	553.856	6	107
9	97.192	2.584	26.249	383.209	9	107
10	49.517	2.652	−57.394	72.190	16	5
11	52.029	6.999	−41.517	254.722	4	11
12	47.000	N/A	N/A	32.900	1	0
13	78.669	8.372	−90.000	1201.706	2	67
14	86.447	6.382	28.966	923.338	3	84
15	104.171	1.891	29.001	296.834	13	122
16	104.732	6.927	45.000	1144.398	3	123
17	124.688	4.262	45.000	716.767	6	165
18	133.551	3.078	45.000	526.740	9	184

The encapsulated samples were clamped in the tensile tester—shown in Figure 6—and elongated at a rate of 0.5 mm/s until failure. At the same time, the DC resistance of the samples was measured using a benchtop multimeter (Keithley 2001), which was triggered by the tensile tester software using a network connection. For this, it was set to auto-ranging with an upper limit of 200 Ω and averaging over 10 measurements—with one measurement per cycle of the 50 Hz power grid—because only the estimated point of failure was of interest, not the precise resistance increase, to determine the point of failure.

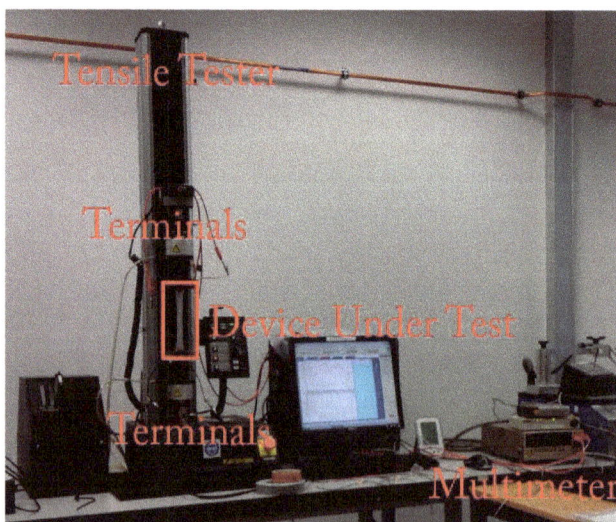

Figure 6. Tensile tester (Instron 5543) with the benchtop multimeter (Keithley 2001) performing the electrical measurements at the right-hand side. The multimeter was triggered over a network connection, and data were recorded using a secondary computer.

3. Results

The meander started out with a resistance of approximately 537 Ω, while it was flat on the carrier board, verified using a zeroed Keysight U1253B multimeter (Keysight Technologies, Santa Rosa, CA, USA). After mounting in the measurement setup, as shown in Figure 3b, this resistance increases slightly to 537.50 Ω, as measured with the Keithley 2400 using a four-wire measurement. The observed resistance as the meander was extended is shown in Figure 7.

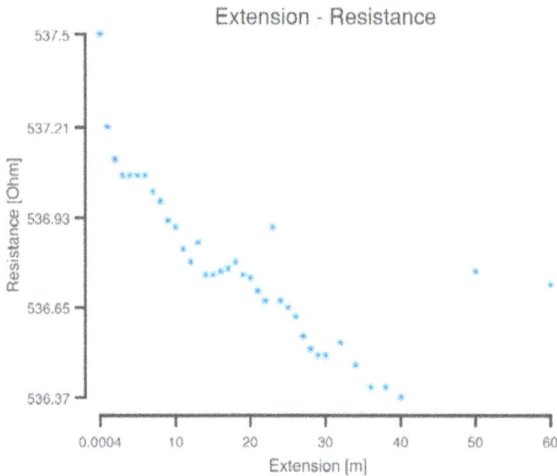

Figure 7. The resistance of the 66.6 m long meander plotted as a function of its extension. A noticeable, but procentual small drop in resistance was observed as the meander was extended, most likely due to the copper folding back on itself at each arc, which would slightly reduce the electric path length when stretching the meander.

A slight drop in resistance from 537.5 Ω to 536.37 Ω was observed during the experiment, contrary to what might be expected from a meander under significant strain. However, this can most likely be explained by the arcs at the end of the U-shaped meanders twisting and folding back on themselves as the meander extends, something slightly visible in Figure 3c. Extension of the meander was stopped at sixty meters because of length limitations of the corridor, limiting the achieved stretchability to 15,000,000%. When attempting to roll the meander on a spool for storage after the experiment, it broke in several places at the points where it twisted during the elongation. More importantly, this demonstrates the design dependence of the term stretchability.

The chosen distance between the start and end point is poorly defined and allows for parlor tricks as demonstrated above. Possible values for this circuit are the width of the cut between the pads (70 μm), the distance between the nearest copper features (400 μm), the distance between the centre of the pads 18.4 mm, the distance between the meander starting points (36.4 mm), or any other ill-devised definition. With each distance leading to its respective extension: 85,714,200%, 14,999,900%, 326,100%, and 164,800%.

The two center lines, A and B, defined in Figure 1a, provide a more realistic length, leading to 2113% and 9288%, respectively. While in this case, center line A provides a more accurate number, the choice between these two center lines can become quite tricky in some instances. For example, using center line B might be justifiable if the radius of 0.86 mm (Figure 2) is increased to 10 mm. In short, defining the initial length against which stretchability is measured has a considerable influence on the result. However, making the *"correct"* choice might not be trivial in some cases.

Needless to say, unless the same definition and measurement guidelines were used, any comparison between these numbers is meaningless. Additionally, this meander will fail whenever the tension on it is released, making it useless for most applications. Clearly, a better definition of stretchability is required to remove ambiguity.

3.1. Alternative Stretchability Metrics

Defeating stretchability as a metric of a stretchable interconnect's capabilities is trivial. As demonstrated above, it only requires creating a long wire using the technology at hand. A useful metric is unbiased and considers the reliability of the technology at hand. Solving this for the case of stretchable electronics seems trivial at first.

For a planar substrate, dividing the maximum achieved elongation by the occupied surface area before elongation would prevent looping the wire around the contact pads, as was done here. In fact, this planar stretchability (PS) should provide a fair assessment of a technology's capabilities at first glance. The maximum PS would depend solely on the minimum feature size and not the design; some small gains could be made by shrinking the contact pads; however, these would be minimal.

$$\text{Planar Stretchability} = \text{Maximum Elongation}/\text{Area} \qquad (1)$$

This, of course, does not account for technologies that use out-of-plane features to create a stretchable interconnect [10,16]. These could create a tower of wire that might take up a relatively large volume while providing no cyclic endurance, like the above circuit—a single elongation might break the interconnect. An easy method to alleviate this is to include the thickness of the circuit by dividing the maximum elongation by the volume of the bounding box around the stretchable interconnect, leading to the volumetric stretchability (VS):

$$\text{Volumetric Stretchability} = \text{Maximum Elongation}/\text{Volume} \qquad (2)$$

To eliminate the effect of the units, each value is normalized to millimeters; alternatively, the unit used could be mentioned next to the variable as a subscript (e.g., PS_{mm}). At first glance, this might result in a constant number for a given technology. However, the design of the interconnect still depends on electrical parameters, such as impedance and maximum current carrying capability, and the mechanics of the encapsulating material.

Using these metrics and looking back at the above example, consider the 60 m extension, 50,905.0 mm^2 surface area, and 3461.5 mm^3 volume (68 μm thickness). The planar stretchability then becomes 1.178 and the volumetric stretchability becomes 17.334. These values are significantly more useful; the required substrate surface area or volume is directly related to the cost to fabricate the device. Additionally, merely dividing the desired final length by the achieved PS or VS will indicate the required area or volume to achieve a given wire length. However, this still would not help determine the actual stretchability a designer or engineer might expect from a circuit—the above circuit is still frightfully unreliable, even though it might score well in a comparison.

Defining reliability as a single number is a troublesome prospect. However, for meanders, a few assumptions are possible. For example, it is possible to postulate the following: "Each directional change between segments of a conductor in a stretchable electrical interconnect where the angle between individual segments exceeds 30 degrees is prone to failure over time". This statement is most certainly wrong in an absolute sense, but provides a straight-forward method to quantify the chance of failure by cutting the stretchable interconnect into segments.

Providing a generic definition for a segment is troublesome; however, using the above definition, a few cases can be defined. First, any sharp corner (<90°) between two straight lines would most definitely be a transition at risk of failure. A second case is an interconnect that consists out of multiple identical elements that are repeated to form an interconnect. The third case, which is close to the above definition, would be a generic catch-all case; consider the tangent along the interconnect its center line. If this tangent changes more than $30°$ versus the tangent at an earlier point on the line this, indicates a transition, and hence a segment. Under ideal circumstances, the first two definitions are used, but these are troublesome to apply to special cases (e.g., fractal meanders).

Consider a segment with a known length, and n segments are required to span this distance. Neglecting the start and end points, the chance of failure for each segment and each transition between individual segments ris $P_1[i]$ and $P_2[i]$, respectively, where i is the cycle number during the cyclic test. Assuming the segments are independent of each other, with an equal chance of failure, and the applied strain is irrelevant, the chance a meander segment will survive a certain stretch cycle becomes the following:

$$P_{Survival}[i] = (1 - P_1[i])(1 - P_2[i]) \tag{3}$$

Hence the chance of failure becomes the following:

$$P_{Failure}[i] = 1 - (1 - P_1[i])(1 - P_2[i]) \tag{4}$$

Taking a few statistical liberties, like assuming the stretch cycles are completely independent of each other, the chance a meander segment breaks after i cycles becomes the following:

$$P_F[i] = 1 - \Pi\,^i_{j=0}\,(1 - P_1[j])(1 - P_2[j]) \tag{5}$$

While this might appear extreme at first glance, the survival chances P_1 and P_2 are close to 1 in most technologies, resulting in a rather small P_F. Assuming each meander segment is independent of its neighbors, the chance of failure in cycle i for n segments then becomes the following:

$$P_F[i, n] = 1 - (\Pi\,^i_{j=0}\,(1 - P_1[j])(1 - P_2[j]))^n \tag{6}$$

Clearly, an increase in n will increase the chance of failure. However, this rudimentary approach fails to take into account the strain the meander might experience; but it does demonstrate the effect of the number of segments or transitions n on meander reliability if these are considered weak points. Introducing this into the stretchability metric is trivial; simply dividing the stretchability by n ought to penalize a technology with many transitions. This leads to the definition of compensated planar stretchability (CPS) and compensated volumetric stretchability (CVS):

$$\text{Compensated Planar Stretchability} = \text{Maximum Elongation}/(n \times \text{Area}) \tag{7}$$

$$\text{Compensated Volumetric Stretchability} = \text{Maximum Elongation}/(n \times \text{Volume}) \tag{8}$$

In both cases, the area and volume are those for the complete meander. The above circuit has a staggering 4033 transitions, which meets the above requirement, slicing the CPS and CVS down to 0.003 and 0.004. For the case of a polymer matrix filled with conductive particles, setting n equal to one is an acceptable choice, unless it is patterned as well, because the likeliness of failure will depend on the material's inherent properties instead of the interconnect design.

An even more generic approach would be to consider the chance of failure occurring per length unit after i cycles $P_{FPL}[i]$. If such a number were available, it could easily be included by substituting n by $(1 - P_{FPL}[i])$. However, this would only provide a momentary comparison point during cycle i. A more generic approach would distill the reliability function $P_{FPL}[i]$ into a single number based on the number of life-cycles a device should survive.

Let $i_{expected}$ and k be the number of stretch-cycles the device should survive during regular use and the percentage of acceptable failures within this period, respectively. While setting k to zero might be an attractive prospect, no economical process achieves a 100% yield. Consider the function f[i] that returns the percentage of failed devices after i cycles; multiplying it by a weighing function and calculating the sum over the entire range of i would return a single number that signifies the reliability.

This weighing function should heavily punish crib deaths (early failures), while not significantly penalizing for failures beyond $i_{expected}$. Based on this, we can propose the weighing function $w[i]$:

$$w[i] = a \times \exp(-b \times i) \wedge \forall i \in \mathbb{N}: w[i] > 0 \wedge a, b \in \mathbb{R}^+_0 \wedge \sum^{+\infty}_{i= i_{expected}} w[i] = 1. \qquad (9)$$

The latter condition in Equation (9) ensures failures after the expect lifetime do not significantly count towards the (un)reliability metric. However, a more reliable technology should still achieve a better result.

Next, factoring in the acceptable percentage of failures is done by the following:

$$a = 1 - k \qquad (10)$$

Once a is known, determining b is trivial by solving the following equation numerically:

$$\exp(b) - 1 = (1 - k) \times \exp(b - b \times i_{expected}) \qquad (11)$$

The reliability metric $R[i_{expected}, k]$ can then be calculated by multiplying the weighing function point-wise with the cumulative failure percentage given by $f[i]$:

$$R[i_{expected}, k] = \sum^{+\infty}_{i = 1} (1 - k) \times \exp(-b \times i) \times f[i] \qquad (12)$$

Using this metric, a higher number will indicate a less reliable technology, meaning it can substitute the compensation factor in Equations (7) and (8).

The exact method used to compare stretchable interconnect technologies and designs should be selected based on the application, available data, and desired outcome. For example, a smart health monitoring patch would only be expected to last a day, while a consumer device in the European Union would have to last for over two years. Sadly, calculating these metrics is infeasible at this time because of insufficient data and will most likely only happen at the point of large-scale industrialization.

3.2. Example Case

Analysing the data from the second test using the above methodology—planar stretchability—demonstrates the usefulness of these modified metrics and their potential pitfalls. Per design, four samples were tested, two with and two without TPU encapsulation. Tables 2 and 3 list the mechanical and electrical measurements, respectively. As mechanical failure, the first point at which the material ruptures was chosen, while electrically, a ten-fold resistance increase versus the starting resistance was used. Because of time uncertainty between the trigger signal being sent and the start of the measurement, one millimeter is deducted from the extension before electrical failure.

The observed failure (Run 2—Design #13) in the mechanical measurement occurred because of a loss of air pressure to the pneumatic grips during the test, releasing the sample. The more common failures during the electrical measurement were caused by a variety of problems between synchronizing the mechanical and electrical measurements. In both cases, the actual achievable elongation values are expected to be less, the reason for this is a combination between the sample slipping and cantilevering in the claws. Measuring the exact length after failure was impossible because of delamination and curling of the material, as shown in Figure 8.

Averaging the values for both the free-standing and encapsulated cases, and separating them into mechanical and electrical failure, the stretchability and planar stretchability were calculated for each value, resulting in Table 4. The considered area when calculating the planar stretchability is the actual width the interconnect takes up on the sample multiplied by 47 mm; the values used are listed in Table 1. The starting fillets were neglected, as these are identical in all cases. However, when comparing technologies, or if they differ between designs, these fillets or other transition structures should be included.

Table 2. Mechanical measurements performed on meanders using a tensile tester.

| | Free Standing | | | | Encapsulated | | | |
| | Run 1 | | Run 2 | | Run 3 | | Run 4 | |
Design	Force [N]	Ext. [mm]	Force [N]	Ext. [mm]	Force [N]	Ext. [mm]	Force [N]	Ext. [mm]
1	7.46	17.3	7.42	17.5	31.65	17.1	37.05	19.1
2	7.66	18.7	7.85	17.9	33.93	18.1	34.02	15.6
3	7.49	16.8	7.51	16.7	34.41	18.0	34.96	16.7
4	6.69	24.7	7.01	29.8	34.71	30.2	35.81	21.1
5	7.41	31.7	7.54	32.2	36.40	25.7	37.20	23.6
6	7.42	31.7	7.29	29.5	33.58	21.6	37.38	24.5
7	8.06	80.2	8.06	79.9	35.85	28.7	35.99	34.3
8	7.03	67.3	7.62	73.8	35.05	34.7	35.52	41.4
9	6.84	64.8	4.67	55.8	36.85	41.2	39.50	45.7
10	6.72	12.7	6.80	13.0	32.00	14.8	28.55	14.4
11	7.82	22.4	7.98	25.1	34.99	21.4	34.47	20.3
12	**7.56**	**14.7**	**7.58**	**14.1**	**34.30**	**16.2**	**31.22**	**14.1**
13	7.60	52.7	Failure	Failure	36.01	27.2	33.58	26.6
14	7.91	70.0	7.69	52.9	38.88	35.7	35.56	29.2
15	5.39	64.4	5.20	64.0	39.08	48.8	37.83	44.5
16	7.94	89.1	7.97	84.8	35.70	31.8	35.62	28.2
17	7.57	107.7	7.57	101.9	39.61	47.3	36.71	38.4
18	6.98	106.0	7.03	108.4	30.01	48.4	37.24	49.5

Table 3. Electrical measurements performed on the meanders during the tensile test.

| | Free Standing | | | | Encapsulated | | | |
| | Run 1 | | Run 2 | | Run 3 | | Run 4 | |
Design	R_{Start} [Ω]	Ext. [mm]	R_{Start} [Ω]	Ext. [mm]	R_{Start} [Ω]	Ext. [mm]	R_{Start} [Ω]	Ext. [mm]
1	0.527	14.5	0.515	16.5	0.521	14.1	0.502	12.2
2	0.509	15.4	0.498	14.3	0.508	11.5	0.496	9.1
3	Failure	Failure	0.470	15.1	0.490	14.0	0.491	13.3
4	0.631	24.1	0.598	23.5	0.613	17.1	0.611	19.7
5	0.618	26.2	0.601	28.4	0.606	19.5	0.582	19.4
6	0.620	25.8	Failure	Failure	0.616	19.8	0.582	21.5
7	0.959	67.4	Failure	Failure	0.879	21.4	0.867	26.9
8	Failure	Failure	0.949	72.8	0.933	26.9	Failure	Failure
9	0.970	64.3	0.979	54.9	0.976	39.5	0.923	45.1
10	0.531	12.2	0.535	12.7	0.531	11.4	0.533	9.7
11	0.515	14.5	Failure	Failure	0.528	15.5	0.532	14.6
12	**0.487**	**7.6**	**0.471**	**11.1**	**0.496**	**9.9**	**0.483**	**12.3**
13	0.820	48.3	Failure	Failure	0.819	21.7	0.779	24.2
14	0.872	63.3	Failure	Failure	0.835	22.4	0.862	20.7
15	1.076	63.9	1.069	63.3	Failure	Failure	1.026	43.9
16	1.046	79.9	1.057	74.0	1.049	28.3	1.044	21.8
17	1.238	104.4	1.206	95.2	1.274	34.1	1.258	27.7
18	1.330	105.4	1.339	107.9	1.413	70.4	1.352	31.5

The straight track reference (Design #12) appears to have a stretchability of 20% to 30%. This high stretchability is not practical because it originates from plastic deformation of the material, combined with the above clamping problems. Additionally, it is non-reversible, limiting its use to one-time deformations. As a result, this should only be considered a baseline measurement.

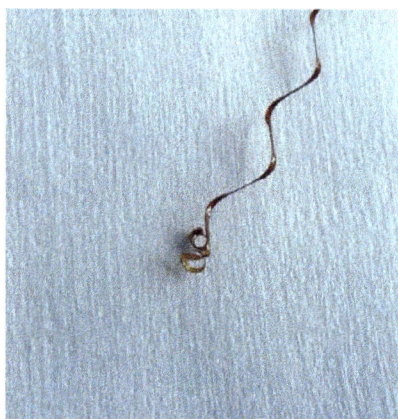

Figure 8. Delamination of the coverlay and curling of the FCB materials after tensile tests performed on the interconnects.

Table 4. Average stretchability and planar stretchability (PS) of the tested interconnects.

| | Mechanical | | | | | | Electrically | | | |
| | Theoretical | | Free Standing | | Encapsulated | | Free Standing | | Encapsulated | |
Design	s [%]	PS [×1000]	s [%]	PS [×1000]	s [%]	PS [×1000]	s [%]	PS [×1000]	s [%]	PS [×1000]
1	8	13	37	62	39	64	33	55	28	47
2	5	18	39	134	36	123	32	108	22	75
3	5	24	36	163	37	169	32	147	29	133
4	32	27	58	49	55	46	51	43	39	33
5	32	51	68	108	52	83	58	92	41	66
6	32	72	65	146	49	110	55	123	44	99
7	107	47	170	76	67	30	143	64	51	23
8	107	90	150	127	81	69	155	131	57	49
9	107	131	128	157	92	113	127	156	90	110
10	5	35	27	178	31	202	26	172	22	146
11	11	20	51	93	44	82	31	57	32	59
12	**0**	**0**	**31**	**438**	**32**	**459**	**20**	**284**	**24**	**337**
13	67	26	112	44	57	22	103	40	49	19
14	84	43	131	67	69	35	135	69	46	23
15	122	193	137	216	99	157	135	214	93	148
16	123	50	185	76	64	26	164	67	53	22
17	165	108	223	146	91	60	212	139	66	43
18	184	164	228	203	104	93	227	203	108	97

Considering the possible deviation caused by plastic deformation and the test (20% to 30%), the experimental stretchability values are closely in line with the theoretical values calculated by considering the meander's centerline—as expected. However, the planar stretchability (PS) values—multiplied times a thousand for the sake of convenience—tell a different story. The clearest example is design #4; while it achieves the same stretchability as designs #5 and #6, it requires significantly more surface area to do so, as illustrated in Figure 9a, making it less attractive from a manufacturing cost perspective.

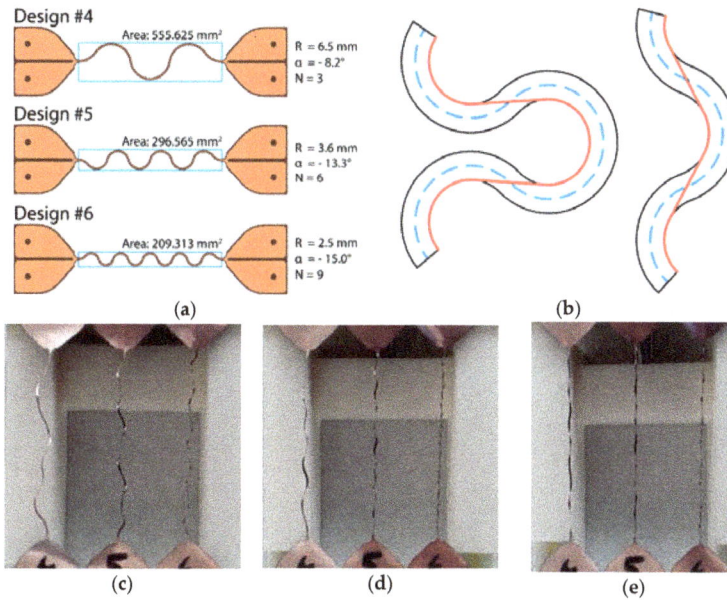

Figure 9. Detailed view of some of the tested samples. (**a**) Parameters and comparison of Designs #4, #5, and #6, illustrating the clear distinctions between designs with similar path lengths. (**b**) The length along the centerline (dashed line) is not the actual length a meander can traverse without undergoing plastic deformation. Instead, the shortest possible distance that traverses the meander track provides a more accurate value. (**c**) Three dummy samples of Design #4, #5, and #6 with a 5 mm extension applied. (**d**) Three dummy samples of Design #4, #5, and #6 with a 10 mm extension applied. (**e**) Three dummy samples of Design #4, #5, and #6 with a 15 mm extension applied.

At the same time, the fallacy of the planar stretchability is magnified by the reference design, which scores a staggering 284 to 438—three to ten times higher than the meanders. This is not surprising, because a straight line is the most efficient way to connect two points on a flat substrate. However, a wider trace would lead to a lower planar stretchability; luckily, the higher stretchability meanders would experience a similar drawback, because they would have to use even wider copper to achieve the same resistance as the much shorter straight trace. However, this method heavily promotes horseshoe-shaped meanders with small radii and a large number of segments because it packs a lot of conductor length per substrate area, even though smaller bending radii might not necessarily be an advantage from a reliability perspective.

The compensated planar stretchability (CPS), listed in Table 5, changes this figure entirely by introducing the number of transitions or segments—available in Table 1. This is, once again, exceptionally clear when comparing designs with similar stretchability, such as Design #4, #5, and #6. While #5 and #6 were previously closely matched because they take up similar substrate surface areas, #4 and #5 now have the advantage because they have less segments. This stands to reason because the increased space between the individual segments would allow more encapsulation material in-between the segments—meaning it can deform more before developing tears in the encapsulation material. An additional reason that Designs #4 and #5 are preferable over #6 can be seen in Figure 9c; at 5 mm elongation, Design #6 is almost entirely stretched, while #4 and #5 still have some headroom. This situation only worsens as the elongation is increased to 10 mm (Figure 9d) and then to 15 mm (Figure 9e). While this might appear counterintuitive at first glance, the reason for this discrepancy can be found in Figure 9b. The length along the meander centerline, as listed in Table 1, is not the actual length a meander can achieve without plastic deformation. Instead, the smallest radii in combination with a tangent running in between these radii provides an accurate number. As a result, the meander with a small number of segments, and hence a lower opening angle α, can potentially span a far longer distance without undergoing plastic deformation.

Instead, the smallest radii in combination with a tangent running in between these radii provides an accurate number. As a result, the meander with a small number of segments, and hence a lower opening angle α, can potentially span a far longer distance without undergoing plastic deformation. For example, when considering the difference between the centerline and this actual length on a per segment basis, Design #4 only loses 4.56% of its length per segment, while this increases to 7.49% and 10.10% for Designs #5 and #6, respectively, explaining the results seen in Figure 9.

Table 5. Compensated planar stretchability (CPS) $\times 1000$.

		Mechanical		Electrical	
	Theoretical	Free Standing	Encapsulated	Free Standing	Encapsulated
Design	CPS [$\times 1000$]	CPS [$\times 1000$]	CPS [$\times 1000$]	CPS [$\times 1000$]	CPS [$\times 1000$]
1	4	21	21	18	16
2	3	22	21	18	13
3	3	18	19	16	15
4	9	16	15	14	11
5	24	18	14	15	11
6	16	16	12	14	11
7	16	25	10	21	8
8	15	21	11	22	8
9	15	17	13	17	12
10	2	11	13	11	9
11	5	23	20	14	15
12	**0**	**438**	**459**	**284**	**337**
13	13	22	11	20	10
14	14	22	12	23	8
15	15	17	12	16	11
16	17	25	9	22	7
17	18	24	10	23	7
18	18	23	10	23	11

The advantages of compensated planar stretchability are also apparent when comparing Designs #16, #17, and #18. While Designs #18 and #17 should significantly exceed the stretchability of Design #16, both their CPS figures are significantly scaled down by introducing the number of segments as a factor. Given that the minimum spacing between two separate meander segments decreases from 5.038 mm to 2.831 mm, and finally to 1.849 mm, respectively, for these three designs, this is a correct figure; especially when considering a larger radius should decrease the strain experienced by the copper.

3.3. Resistance Measurements

DC resistance measurements are the golden standard to electrically verify stretchable electronics. The general idea consists of proving the prowess of the technology or design at its intended application; that is, passing an electrical signal. However, many measurements downplay the actual effects observed or are fundamentally flawed from the start by failing to consider realistic use conditions.

The above circuit was tested using a 1 mA current, a typical test current for resistance measurement to place the returned voltage in the multimeter's 2.1 V measurement range. However, if the function of the device was to carry a large current (e.g., 10 A), the performed measurement would be pointless because the heating might significantly affect the mechanical reliability of the circuit. Carrying large currents will pose different challenges as opposed to, for example, capacitive sensing, the measurement should use a test current close to the one encountered in the intended application whenever possible. This is especially important when considering the failure mode of conductors can change dramatically as the width and thickness increases, as commonly seen with flexible circuit boards, meaning a small dimensional change to improve electrical performance might cause significant issues mechanically.

Under any condition, the first step is correctly measuring the observed resistance. Three circuit topologies, shown in Figure 10, are in everyday use when measuring the resistance of an electrical interconnect. The first topology (Figure 10a) is equivalent to placing multimeter probes on the contact pads and performing a two-point measurement. Here, both the contact and lead resistance come into full effect. The next possible topology is the quasi four-wire approach (Figure 10b), where separate wires are used for driving the measurement current and sensing the voltage, but they are connected before contacting the device-under-test (DUT). As a result, the contact resistance is still present. The final case, a true four-wire measurement (Figure 10c), is the only correct method in most cases [17,18].

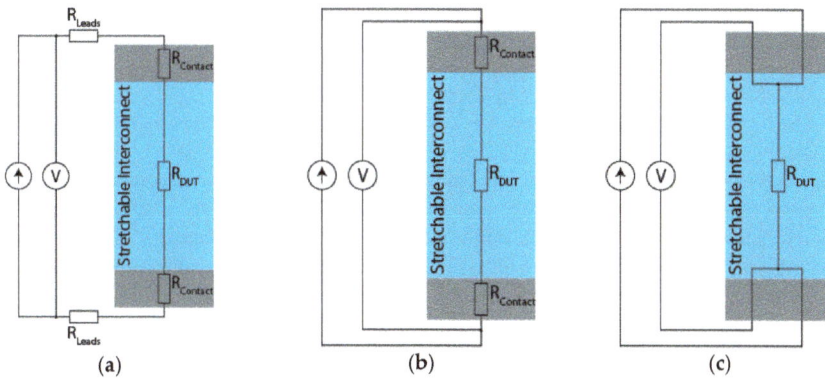

Figure 10. Measurement setups for resistance measurement: (**a**) two-wire setup affected by lead and contact resistance; (**b**) quasi four-wire setup affected by the contact resistance; and (**c**) true four-wire setup that eliminates the effect of lead and contact resistance.

The method in Figure 10a is only suitable for quick process verification, secondary checks, or when dealing with resistances of multiple kilo-ohms. For the tested sixty-meter meander, Figure 10b presents the minimum acceptable measurement. The movement of the test leads, and especially the coiling on the reel, might significantly affect the resistance of the test leads during the measurement. However, during movement, the contact resistance can change dramatically, making the true four-wire measurement the only correct method.

To understand the importance, it is worth looking at ballpark figures. Consider a stretchable interconnect in an undefined technology; this hypothetical copper-based interconnect has a resistance of 300 mΩ. A two-wire measurement (Figure 10a) is performed on the interconnect while it is stretched, hoping to monitor the formation of micro-cracks over time. The test leads have a resistance of 100 mΩ each, while the nickel-plated probes have a contact resistance of 25 mΩ with interconnect. Before stretching the interconnect, the multimeter will see 550 mΩ, which, for example, increases to 700 mΩ when the interconnect is fully extended—or a 27.3% increase. While actually, the resistance increase the meander experienced was 50%, a significantly different result.

The next measurement, performed by a slightly more skilled operator, uses a zeroed multimeter—meaning the probes were placed on the same contact pad to make a reference measurement. This action will indeed subtract 250 mΩ from the measured resistance and provide a 50% figure in theory. However, in practice, the measurement is still flawed and the formation of small micro-cracks will still be lost to the measurement noise and error. The movement during stretching will slightly change the resistance of the test leads, and shifting of the contacts will result in a slight, but noticeable alteration of the contact resistance because it is a function of pressure, surface condition, and location [19–21]. The magnitude of this effect is difficult to estimate as it heavily depends on the surface roughness and type of probe. However, it is safe to say it will create an uncertainty in the milliohm range or larger (e.g., ±5 mΩ). Additionally, the multimeter might not be in its optimal measurement range; the software will dutifully subtract the 250 mΩ and present a 150 mΩ in ideal circumstances. However, if the measurement ranges were, for example, 500 mΩ and 5 Ω with a resolution of 1 mΩ and 10 mΩ, respectively, the measurement uncertainty would increase tenfold for this scenario—because the multimeter is now operating in its 500 mΩ range. This means that it is crucial to consider the measurement range the multimeter is in, and not blindly base the result on the displayed value, especially when dealing with low-cost handheld units.

Of course, not every effect comes into play in every situation; therefore, it is essential to understand the magnitude of the problem for a specific situation. Micro-cracks are initially virtually impossible to detect electrically; this is easy to prove using simple electrical theory. The resistance of a rectangular conductor at DC depends on its resistivity ρ, length L, thickness t, and width w [22]:

$$R = \rho \times L / (t \times w). \tag{13}$$

Consider a copper trace with a thickness of 18 μm and width of 100 μm at 20 °C, it will have a resistivity of 16.8 nΩm [23,24]. If the slice in which the micro-crack occurs has a length of 1 μm, it will start with a resistance of 9.3 μΩ. To add 1 μΩ of resistance, the crack has to propagate 9.7 μm into the trace, and another 7.95 μm to add another 1 μΩ. In short, detecting micro-cracks early requires exact high-resolution resistance measurements. This also explains why failures might appear suddenly; the trace might have broken most of the way before the actual resistance increase becomes noticeable. Of course, detecting a single micro-crack tends to be impossible under most circumstances, luckily, they tend to occur in large numbers. However, based on this, we can conclude analyzing the mechanical condition of a circuit based on electrical measurements is a precarious method unless the measurement limitations are considered.

What is also interesting to study is the effect of this localized resistance increase when carrying larger currents. Consider, for example, the case of a constant current LED driver supplying 100 mA through the above 18 μm by 100 μm long interconnect. Assuming the interconnect spans a distance of 200 mm, it will have a resistance of 1.867 Ω and will dissipate approximately 18.7 mW over its entire length—a trivial amount not worth mentioning for a highly conductive material like copper. However, at the location of micro-cracks, things can turn for the worse quickly; a 1-μm long slice will dissipate only 93.33 nW, but this goes up to 933.3 nW for a 90-μm wide crack. Considering this slice weighs only 16.13 ng and copper has a specific heat capacity of 384.4 J/(kg K) [24], it requires only 620 nJ to increase its temperature by 100 °C—or less than one second at this current assuming no thermal energy is conducted to the surrounding material. While conduction to the surroundings will dampen the temperature rise, it does point out that thermal degradation of the encapsulating material, which supports the interconnect, can occur at the site of failing conductors. It is not far-fetched that this might significantly aggravate the situation and lead to earlier failures depending on the encapsulation material.

Another effect worth considering when measuring small resistances is the Seebeck effect, meaning any metal-to-metal junction starts behaving as a thermocouple [17]. In most measurement setups, this effect is trivial and readily eliminated by using a parallel measurement path on both sides of the sample, because the entire setup is at the same temperature. Alternatively, specific alloys can be used on the contacts to limit this effect further—usually labelled as "low thermal electromotive force (EMF)" by the manufacturer. However, heating caused by the circuit or mechanical movement can upset this isothermal environment and result in the introduction of large thermal voltages.

Consider a typical nickel-plated probe; the copper–nickel Seebeck coefficient is 10 μV/K for pure copper and nickel [25]. A temperature gradient of 10 °C will then lead to a thermal voltage of 100 μV. Using the initial example once more, a source meter sends a 100 mA measurement current through 300 mΩ, which results in a voltage of 30 mV over the sense leads in ideal conditions. The 100 μV offset by the thermal voltage skews this measurement to 301 mΩ, a relatively small error. The problem stems from the fact that many handheld multimeters are incapable of providing high test currents, unlike source meters or high-end bench meters, which will happily provide higher test currents. For example, the Keysight U1253B uses a current of 1.04 mA, which would result in a 312 μV output, adding 100 μV onto that results in a 400 mΩ value on the display (10 mΩ resolution). An even worse situation exists when considering copper–copper oxide contacts, which can exhibit Seebeck coefficients of 1 mV/K [17]. For this reason, choosing the correct probe for the intended application is of vital importance when measuring small resistances.

A final consideration is the time it takes the multimeter to perform the measurement. This measurement is rarely instant because the sample-and-hold circuit requires time to charge a capacitor [26]. A changing resistance during this interval—called the aperture time—is a potential cause for error, meaning movement of either the probes or the sample can introduce an unintended bias to the measurement [18]. This is exceptionally clear when attempting to eliminate the noise of the power grid by using aperture times that are multiples of 20 ms or 16.7 ms to average one or more power cycles. Ideally, the mechanical measurement stops, triggers the multimeter, and waits for it to complete a measurement before continuing the test.

In short, measuring small resistance changes requires a correctly configured measurement setup. The most substantial errors can be avoided by using a four-wire measurement with appropriate test currents to emulate self-heating. Samples should have clean electrical contacts at the same temperature, and the measurement instrument has to be configured to the correct range with a suitable test current.

4. Discussion

Stretchability is commonly used to describe the performance of a stretchable electronic interconnect. However, as demonstrated above, it is an unreliable metric for the performance of a technology because of its design dependence. A series of alternative metrics taking into account the area, volume, and reliability of the interconnect were proposed to limit the influence of the design. Additionally, the considerations for accurate resistance measurement were highlighted using simple examples, providing a straightforward way to estimate their importance for typical stretchable interconnect measurements.

Supplementary Materials: The following are available online at http://www.mdpi.com/2072-666X/9/8/382/s1, Figure S1: AutoCAD circuit design file for the 60-m meander, Figure S2: AutoCAD circuit design file for the meanders tested using the tensile tester.

Author Contributions: Conceptualization, B.P. and J.V.; Investigation, B.P.; Methodology, B.P.; Visualization, B.P.; Writing—original draft, B.P.; Writing—review & editing, B.P., F.B., and J.V.

Funding: This research received no external funding.

Acknowledgments: The authors would like to thank Sheila Dunphy, Kristof Dhaenens, and Steven Van Put for their assistance in fabricating the samples.

Conflicts of Interest: The authors declare no conflict of interest.

References

1. Wagner, S.; Bauer, S. Materials for stretchable electronics. *MRS Bull.* **2012**, *37*, 207–213. [CrossRef]
2. Rogers, J.A.; Someya, T.; Huang, Y. Materials and mechanics for stretchable electronics. *Science* **2010**, *327*, 1603–1607. [CrossRef] [PubMed]
3. Isaac, S.; Leah, S. Conductor Reel. U.S. Patent 2,031,434, 18 Feburary 1936.
4. Coombs, C.F.; Holden, H. *Printed Circuits Handbook*, 7th ed.; McGraw-Hill Education: New York, NY, USA, 2016.

5. Someya, T. *Stretchable Electronics*; Wiley: Hoboken, NJ, USA, 2013.
6. Gray, D.S.; Tien, J.; Chen, C.S. High-conductivity elastomeric electronics. *Adv. Mater.* **2004**, *16*, 393–397. [CrossRef]
7. Plovie, B.; Yang, Y.; Joren, G.; Sheila, D.; Kristof, D.; Steven, V.P.; Björn, V.; Thomas, V.; Frederick, B.; Jan, V. Arbitrarily shaped 2.5d circuits using stretchable interconnects embedded in thermoplastic polymers. *Adv. Eng. Mater.* **2017**, *19*, 1700032. [CrossRef]
8. Vanfleteren, J.; Löher, T.; Gonzalez, M.; Bossuyt, F.; Vervust, T.; Wolf, I.D.; Jablonski, M. Scb and smi: Two stretchable circuit technologies, based on standard printed circuit board processes. *Circuit World* **2012**, *38*, 232–242. [CrossRef]
9. Yang, Y.; Thomas, V.; Sheila, D.; Steven, P.; Bjorn, V.; Kristof, D.; Lieven, D.; Lothar, M.; Linde, V.; Tom, M.; et al. 3d multifunctional composites based on large-area stretchable circuit with thermoforming technology. *Adv. Electron. Mater.* **2018**. [CrossRef]
10. Sun, Y.; Choi, W.M.; Jiang, H.; Huang, Y.Y.; Rogers, J.A. Controlled buckling of semiconductor nanoribbons for stretchable electronics. *Nat. Nanotechnol.* **2006**, *1*, 201. [CrossRef] [PubMed]
11. Carta, R.; Jourand, P.; Hermans, B.; Thoné, J.; Brosteaux, D.; Vervust, T.; Bossuyt, F.; Axisa, F.; Vanfleteren, J.; Puers, R. Design and implementation of advanced systems in a flexible-stretchable technology for biomedical applications. *Sens. Actuators A Phys.* **2009**, *156*, 79–87. [CrossRef]
12. Kubo, M.; Li, X.; Kim, C.; Hashimoto, M.; Wiley, B.J.; Ham, D.; Whitesides, G.M. Stretchable microfluidic radiofrequency antennas. *Adv. Mater.* **2010**, *22*, 2749–2752. [CrossRef] [PubMed]
13. Hsu, Y.-Y.; Gonzalez, M.; Bossuyt, F.; Axisa, F.; Vanfleteren, J.; De Wolf, I. The effects of encapsulation on deformation behavior and failure mechanisms of stretchable interconnects. *Thin Solid Films* **2011**, *519*, 2225–2234. [CrossRef]
14. Kwon, S.; Cho, H.W.; Gwon, G.; Kim, H.; Sung, B.J. Effects of shape and flexibility of conductive fillers in nanocomposites on percolating network formation and electrical conductivity. *Phys. Rev. E* **2016**, *93*, 032501. [CrossRef] [PubMed]
15. Tao, C.; Yizhou, Z.; Wen-Yong, L.; Wei, H. Stretchable thin-film electrodes for flexible electronics with high deformability and stretchability. *Adv. Mater.* **2015**, *27*, 3349–3376.
16. Kim, D.H.; Xiao, J.; Song, J.; Huang, Y.; Rogers, J.A. Stretchable, curvilinear electronics based on inorganic materials. *Adv. Mater.* **2010**, *22*, 2108–2124. [CrossRef] [PubMed]
17. Keithley. *Low Level Measurements Handbook*, 7th ed.; Tektronix: Beaverton, OR, USA, 2013; p. 244.
18. Horowitz, P.; Hill, W. *The Art of Electronics*; Cambridge University Press: Cambridge, UK, 2015.
19. Kogut, L.; Komvopoulos, K. Electrical contact resistance theory for conductive rough surfaces. *J. Appl. Phys.* **2003**, *94*, 3153–3162. [CrossRef]
20. Hyun, C.; Park, H.; Hahm, S. *An Analysis of Probing Cres in Gold Bumping Pad Using Automatic Test Equipment*; IEEE SW Test Workshop: San Diego, CA, USA, 2012.
21. Timsit, S. Electrical Contact Resistance: Properties of Stationary Interfaces, Electrical Contacts–1998. In Proceedings of the Forty-Fourth IEEE Holm Conference on Electrical Contacts (Cat. No.98CB36238), Arlington, VA, USA, 26–28 October 1998; pp. 1–19.
22. Kumar, N. *Comprehensive Physics Xii*; Laxmi Publications: New Delhi, India, 2004.
23. Matula, R.A. Electrical resistivity of copper, gold, palladium, and silver. *J. Phys. Chem. Ref. Data* **1979**, *8*, 1147–1298. [CrossRef]
24. Kaye, G.W.C.; Laby, T.H. *Tables of Physical and Chemical Constants*; Longman: Harlow, UK, 1995.
25. Kidd, M. Watch out for those thermoelectric voltages! *Int. J. Metrol.* **2012**, *19*, 18–21.
26. Analog Devices Inc., E; Zumbahlen, H. *Linear Circuit Design Handbook*; Elsevier Science: New York, NY, USA, 2011.

micromachines

MDPI

Article

Flexible Thermo-Optic Variable Attenuator based on Long-Range Surface Plasmon-Polariton Waveguides

Jie Tang [1,2], Yi-Ran Liu [2,3], Li-Jiang Zhang [3], Xing-Chang Fu [3], Xiao-Mei Xue [1,2], Guang Qian [1], Ning Zhao [2,3] and Tong Zhang [1,2,3,*]

[1] Key Laboratory of Micro-Inertial Instrument and Advanced Navigation Technology, Ministry of Education, and School of Instrument Science and Engineering, Southeast University, Nanjing 210096, China; tangjieck@126.com (J.T.); xiaomei_xue2011@163.com (X.-M.X.); chinaqgll@163.com (G.Q.)
[2] Suzhou Key Laboratory of Metal Nano-Optoelectronic Technology, Suzhou Research Institute of Southeast University, Suzhou 215123, China; duduranla@163.com (Y.-R.L.); njzhao88@163.com (N.Z.)
[3] Joint International Research Laboratory of Information Display and Visualization, School of Electronic Science and Engineering, Southeast University, Nanjing 210096, China; ljzhv@163.com (L.-J.Z.); pasf365@163.com (X.-C.F.)
* Correspondence: tzhang@seu.edu.cn; Tel.: +86-025-8379-2449

Received: 28 April 2018; Accepted: 24 July 2018; Published: 26 July 2018

Abstract: A flexible thermo-optic variable attenuator based on long-range surface plasmon-polariton (LRSPP) waveguide for microwave photonic application was investigated. Low-loss polymer materials and high-quality silver strip were served as cladding layers and core layer of the LRSPP waveguide, respectively. By using finite element method (FEM), the thermal distribution and the optical field distribution have been carefully optimized. The fabricated device was characterized by end-fire excitation with a 1550 nm laser. The transmission performance of high-speed data and microwave modulated optical signal was measured while using a broadband microwave photonics link. The results indicated that the propagation loss of the LRSPP waveguide was about 1.92 dB/cm. The maximum attenuation of optical signal was about 28 dB at a driving voltage of 4.17 V, and the variable attenuation of microwave signals was obviously observed by applying different driving voltage to the heater. This flexible plasmonic variable attenuator is promising for chip-scale interconnection in high-density photonic integrated circuits and data transmission and amplitude control in microwave photonic systems.

Keywords: variable optical attenuator (VOA); surface plasmon-polariton (SPP); microwave photonics

1. Introduction

The flexible electronics and optical devices are promising for smart wearable devices, large capacity data communications, integrated optical sensing elements, and integrated optoelectronics [1–3]. Recently, the demands for high speed and large capacity interconnections in the optical communications and chip-to-chip signal transmissions increase rapidly. Flexible optical interconnection technology is an attractive solution for these applications, owing to the large bandwidth and the flexibility of the optical signals [4,5].

The variable optical attenuator (VOA) is an important optical component that is widely used in optical communication systems [6,7] and optical signal processing in microwave photonics [8,9]. Especially in microwave photonic systems, microwave signals are transmitted and processed in optical domain and the VOAs are always used to adjust and equalize the power of the optical signals in different parallel channels. With the development of integrated optics, many chip-scale integrated microwave photonic circuits that are used for signal processing and data transceiving have been realized [10,11]. In the future, large-scale multifunctional integrated optical systems will be implemented by using chip-to-chip hybrid interconnections. Hence, the VOAs with wide attenuation range, high-density integration, and flexible bending are necessary for parallel multi-channel optical interconnections.

In recent years, many kinds of VOAs with different structure and mechanism have been reported, such as radiation loss tuning in the S-bend waveguide [12], phase adjusting in Mach–Zehnder interferometers (MZIs) [13,14], optical evanescent field absorption in the waveguide [15], and light mode tuning in a multimode interference device (MMI) [16–20], etc. Although these works above present excellent performance in different aspects, such as low-power, low insertion loss, large attenuation range, but still not adequate to meet the requirements of sub-wavelength optical manipulation as well as transmission in higher integrated circuits in the future for the optical diffraction limit [12,21].

The surface plasmon-polariton (SPP), which is a transverse magnetic (TM) polarized plasmon mode that exists in the interface between a metal and a dielectric, has been shown to transmit optical signals beyond the diffraction limit [22–24]. It is very advantageous for high-density integrated optical systems [21]. With a symmetrical dielectric cladding and thin film core, the surface plasmon modes that are associated with the upper and lower of the metal-dielectric interfaces couple and form a low-loss symmetric mode, known as the long-range SPP (LRSPP) [25]. In comparison with the asymmetric short-range SPP (SRSPP) mode, the LRSPP mode has a large propagation length due to its low propagation loss and it can be applied to chip-to-chip interconnections. But, the SRSPP mode generally has a small mode size and it is suitable for compact integrated optic devices and hybrid integrated chips [26,27]. The SPP waveguides have been employed in various optical devices, such as VOAs [28], filters [29], switches [30], and modulators [31–33] for the refractive indices of the claddings can be adjusted by thermo-optic or electro-optic effect. The LRSPP waveguide consisting of dielectric claddings with similar refractive indices and the metal thin film is extremely sensitive to the symmetry of refractive indices. The mode cut-off of LRSPP in the metal strip can be acquired with an asymmetry in the refractive indices of the dielectric layers above and below the thin metal strip. For a slight asymmetry, the LRSPP mode is still supported, while if the asymmetry is too large the LRSPP mode becomes cut-off and no low loss modes are supported, resulting in asymmetric SRSPP mode with high propagation loss. Light input into the structure radiates away from the metal strip core. Besides, the mode size of the LRSPP waveguide can be adjusted by changing the thickness and width of the stripe core to obtain a low coupling loss with other kinds of waveguide devices or fibers.

Based on the sensitivity to interface property of the LRSPP waveguide mentioned above, several VOAs have been proposed and investigated. For example, a LRSPP VOA that is based on electro-optic controllable liquid crystals and polymers is investigated and it exhibits a good performance of low power, high extinction ratio, and low insertion loss [34]. For both polarization use, a plasmonic nanowire-based thermo-optic VOA with a cross-section of 250 nm × 250 nm shows a low polarization dependent loss of ±2.5 dB with a compact footprint of 1 mm [35]. Another thermo-optic LRSPP modulator with a length of 1 cm is investigated at telecom wavelengths and it shows low driving power and high extinction ratio [36]. However, most of these VOAs use metal strip core as the heater, which will limit the physical lifetime of the device because the thin metal strip core may be wrinkled and damaged at a high temperature.

In this paper, we design a flexible VOA with a structure consisting of LRSPP waveguide and a thermal heater. By using a simple spin-coating, photolithography, and wet etching process, the device has been fabricated and then characterized at a wavelength of 1.55 μm. The performance of optical attenuation and high speed data transmission are measured.

2. Device Design and Simulation

The schematic diagram of the VOA based on LRSPP waveguide is shown in Figure 1. The bottom left inset is the cross-section of the device. It consists of a gold (Au) layer, lower polymer cladding, thin metal strip core, upper polymer cladding and a heater aligned on top of the upper cladding.

The thicknesses of the upper and lower polymer cladding layers are both 15 μm. The silver (Ag) strip core of the LRSPP waveguide is carefully designed with an optimized parameter and the width and thickness is 4 μm and 12 nm, respectively. The heater is made of aluminum (Al). Its cross-section dimension is 10 μm wide and 150 nm thick and the length is 10 mm. The refractive indices of the polymer claddings are both 1.45 and that of the Ag strip core is 0.15 + 11.38i at 1550 nm. The calculated optical field distribution by the finite element method (FEM) is shown in the top right inset of Figure 1. It indicates that the optical field is symmetrically distributed in the interface between the claddings and the strip core, and most of the energy is distributed in the claddings next to the interface, which ensures a long-range propagation mode.

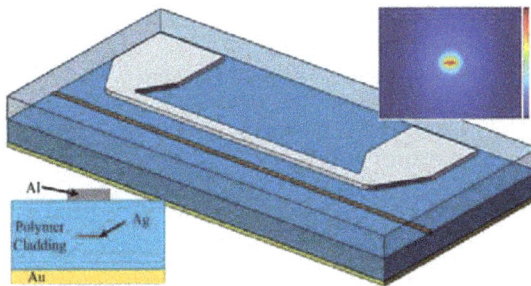

Figure 1. The schematic diagram of the variable optical attenuator (VOA). The bottom left inset is the cross-section and the top right inset is the optical field distribution of the long-range surface plasmon-polariton (LRSPP) waveguide.

The principle of the VOA is based on changing the radiation loss of the LRSPP waveguide resulting from the refractive index difference between the upper and lower claddings. When a voltage is applied to the heater, thermal energy being produced by resistive heating is transferred to the upper cladding, Ag strip, and lower cladding in turn. Then, a temperature gradient is established by the thermal energy and it increases with the applied voltage, as shown in Figure 2. The temperature gradient subsequently results in the refractive index difference between the upper and lower cladding due to the thermo-optic effect. Finally, the asymmetry of refractive indices results in the radiation loss of the LRSPP mode. Therefore, the optical output power of the device can be controlled by changing the temperature gradient by applying different voltages to the heater. The thermo-optic coefficient of the polymer is $-1.758 \times 10^{-4}/°C$. The negative thermo-optic coefficient results in the refractive index decrease with the increase in temperature.

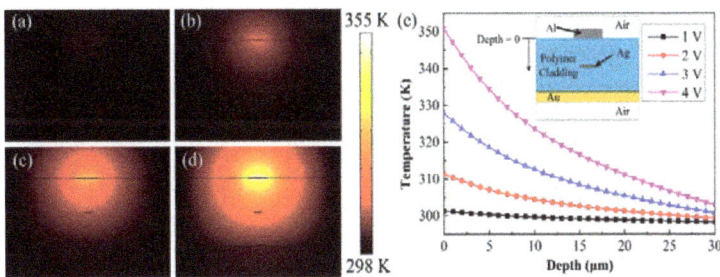

Figure 2. The calculated thermal distribution with applied voltage of (**a**) 1 V, (**b**) 2 V, (**c**) 3 V, (**d**) 4 V to the heater. (**e**) Temperature gradient in the depth of the device at different applied voltage.

According to the dimension of the heater, we can calculate resistance of the heater. By using FEM, we calculate the thermal distribution in the device while applying different voltages to the heater, as shown in Figure 2a–d. The corresponding temperature distribution is shown in Figure 2e. It shows a clear asymmetry distribution of the temperature in the upper and lower claddings when the voltage is applied. Temperature gradient also increases rapidly with the increasing applied voltage (see Figure 2e), which will result in a larger refractive index gradient and eventually enlarge the radiation loss of the LRSPP mode. Therefore, the propagation loss of the device can be adjusted by changing the applied voltage to the heater.

3. Experiments

Figure 3 presents the fabrication process of the flexible thermo-optic VOA based on LRSPP waveguide. Firstly, a layer of 40 nm thick Au was thermally evaporated on the silicon wafer as an adhesion layer, which made it easy to peel off the flexible device from the wafer in the last step. Then, a layer of 15 μm thick ultraviolet (UV) cured epoxy resin was spin-coated on the Au layer as the lower polymer cladding. After a subsequent bake at 160 °C for 30 min, a layer of 2.5 μm thick negative photoresist was spin-coated and baked at 110 °C for 3 min. With a UV photolithography and development, the photoresist was patterned to fabricate a 4 μm wide strip groove. The electron beam evaporation method was used to deposit high-quality Ag thin metal film of 12 nm thick onto the negative photoresist layer. In the evaporation process, a high-purity (>99.999%) solid Ag and a very low deposit speed of 0.02 nm/s in a high vacuum of 1×10^{-4} Pa were used to ensure a high-quality Ag film. After that, the patterned negative photoresist layer was lift-off and a 4 μm wide Ag strip was left over to form the core layer of LRSPP waveguide. Then, another layer of 15 μm thick UV-cured epoxy resin was spin-coated and cured as upper cladding. To fabricate the heater, a layer of 150 nm Al film was thermally evaporated. Then, a layer of positive photoresist was spin-coated and patterned by UV photolithography. After that, wet etching was used to fabricate the Al heater with sodium hydroxide (NaOH) solution. Finally, the flexible device was lift-off from the silicon wafer. After the fabrication process above, the two end faces of the device were polished to minimize the coupling loss between the fiber arrays and the waveguide.

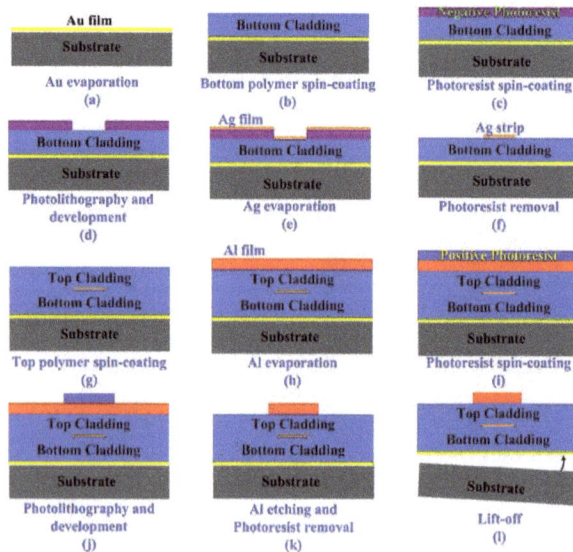

Figure 3. Fabrication process of the flexible LRSPP based VOA.

The schematic diagram of the measurement setup for the thermo-optic VOA is shown in Figure 4. To study the transmission characteristics of high frequency microwave signals, we constructed a broadband microwave photonics link with a bandwidth over 25 GHz. Vector network analyzer (VNA) with a bandwidth of 50 GHz was used to measure the transmission characteristics of the high frequency microwave signals. The light at 1550 nm from a distributed feedback (DFB) laser source was modulated by high frequency microwave signals via a lithium niobate (LiNbO$_3$) electro-optic modulator (EOM). The modulated light was perpendicularly polarized to the waveguide plane by passing through a polarization controller (PC) and then coupled into the waveguide to excite LRSPP mode through a standard single mode fiber (SMF) array. To measure the optical attenuation characteristics of the VOA, a direct-current (DC) power supply was used to apply different voltages to the heater through two probes. In the output port of the device, a standard SMF array was used in order to couple the output light signal into an optical power meter to measure the propagation loss of the LRSPP waveguide and the optical attenuation characteristics of the VOA. Alternatively, the output light was also coupled into a broadband photodetector to demodulate the high frequency microwave signals. Herein, we used an Erbium-doped fiber amplifier (EDFA) with a gain of 20 dB to compensate for the insertion loss of the LRSPP waveguide-based VOA. Meanwhile, a microwave power amplifier (PA) with a gain of 19 dB was employed to compensate the total loss of the microwave photonic link. The fabricated thermo-optic VOA under test is shown in the top left inset of Figure 5.

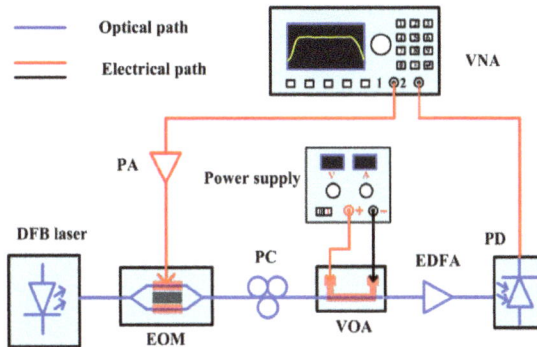

Figure 4. The measurement setup for the test of the optical attenuation and microwave transmission characteristics.

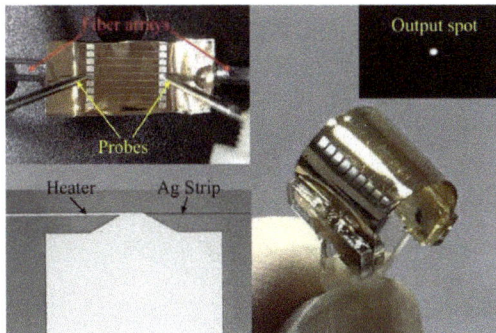

Figure 5. The photograph of the curved flexible thermo-optic VOA. Top left inset is the device under test and the driving voltage is applied through two probes. Top right inset is the near-field output spot of the LRSPP waveguide. Bottom left inset is the photograph taken by an optical microscope.

4. Results and Discussion

The output light of LRSPP waveguide was observed before packing with the fiber array by using an infrared (IR) charge-coupled device (CCD). The near-field output spot of the device is shown in the top right inset of Figure 5. The bright spot shows that the fabricated LRSPP waveguide has a good performance in terms of light propagation. By coupling the output light into the optical power meter, the propagation loss of the LRSPP waveguide is measured at about 1.92 dB/cm by cut-back method, as shown in Figure 6. The results show that the propagation loss is at least two times lower than other previously reported flexible LRSPP waveguide [37,38]. It also indicates that the quality of the fabricated LRSPP waveguide is very high. In order to investigate the performance of the optical attenuation, the voltage is applied to the heater. As the applied voltage increases, the optical attenuation can be observed clearly, as shown in Figure 7. The measured results show that a maximum optical attenuation of over 28 dB is acquired at a driving voltage of 4.17 V. At the same time, we can see that the optical attenuation is inconspicuous when the applied voltage is lower than about 2.5 V. While the applied voltage is larger than 2.5 V, the output optical power decreases rapidly. The reason of this phenomenon can be attributed to the fact that the LRSPP mode still exists in spite of a slight asymmetry of the refractive index that is produced by a relatively low driving voltage. When a larger voltage is applied, the asymmetry is too large and the LRSPP mode becomes cut-off. In other words, the SPP mode changes from nonradiative mode to radiative mode in the case of a large asymmetry of the refractive index.

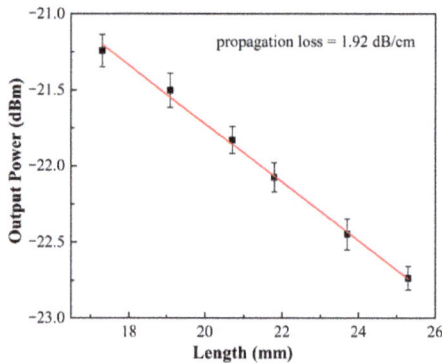

Figure 6. Measured propagation loss of the LRSPP waveguide by cut-back method.

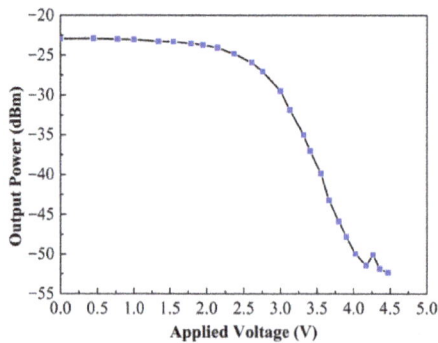

Figure 7. Measured optical output power of the thermo-optic VOA as a function of applied voltage to the heater.

By using the microwave photonics link and the VNA, we can observe the transmission and attenuation characteristics of high frequency microwave signals. The microwave attenuation can be attributed to the decrease of the optical light coupled into the photodetector. According to the measured results of the optical attenuation above, the maximum optical attenuation of 28 dB can result in a maximum microwave attenuation of 56 dB for the square relation between optical loss and microwave loss. The measured microwave transmission and attenuation characteristics under different applied voltage are shown in Figure 8. In the experiment, we carefully adjust the voltage in order to acquire each microwave attenuation interval of 5 dB. Similar to the optical attenuation characteristics, the microwave attenuation increases with the increase of applied voltage. The microwave attenuation is also inconspicuous when the applied voltage is lower than about 2.5 V. Only 5 dB attenuation is acquired when the applied voltage increases from 0 V to 2.51 V. While the applied voltage is larger than 2.5 V, the microwave attenuation increases rapidly. Over 50 dB attenuation is acquired as the voltage increases from 2.51 V to 4.05 V. As the applied voltage increases, more noise appears in the transmission curve. The reason is that the optical power reached in the photodetector is too low to demodulate the high frequency microwave signals. In spite of the noise in the curve, a maximum attenuation of about 55 dB is acquired.

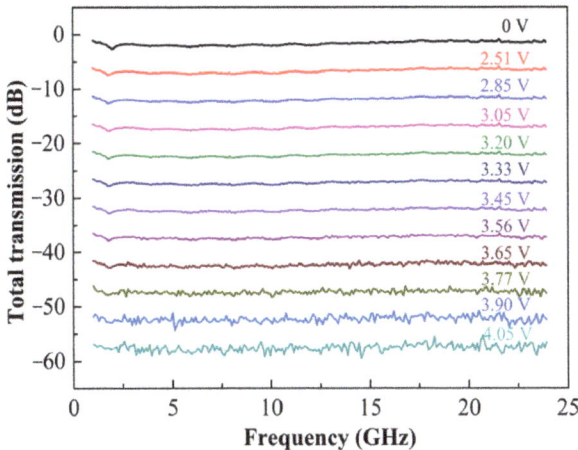

Figure 8. The total microwave transmission characteristics of the microwave photonics link at different applied voltage to the heater.

After the characterization of optical propagation loss and microwave transmission performance, we also carried out a data transmission experiment with the flexible LRSPP waveguide by using a bit-error-rate (BER) test system (N4903B, Agilent Technology, Santa Clara, CA, USA). Figure 9 shows the measured eye diagrams of the total microwave photonics link. A modulated light with 3 Gb/s, 6 Gb/s, 10 Gb/s, and 12.5 Gb/s pseudorandom-binary sequence (PRBS) ($2^{31} - 1$) generated by a pulse-pattern-generator (PPG) was transmitted through the flexible LRSPP waveguide. As shown in Figure 9, the eye is well open at a different transmission speed. The BER at 12.5 Gb/s was measured about 1.13×10^{-10}. The results indicate that the flexible LRSPP waveguide is suitable for high speed optical interconnections.

Evidently, the fabricated devices still suffer from too high insertion loss to meet practical applications. In comparison with the other low power VOAs [39–41], the power consumption is a little larger as well. It is mainly due to the thick polymer cladding that extends the distance of thermal diffusion. However, a good performance of the optical and microwave attenuation has been acquired. Furthermore, the transmission of a light signal modulated by high frequency microwave signals through the flexible LRSPP waveguide-based VOA has been tested in spite of obstacles of large insertion loss, wave vector difference, and dispersion of the LRSPP mode. It is a potential solution to realize the flexible chip-scale interconnections and power equalization in the optical systems. The problem of large insertion loss can be solved by further optimizing the coupling loss between fiber and waveguide or using gain media doped polymer to reduce the propagation loss in the future.

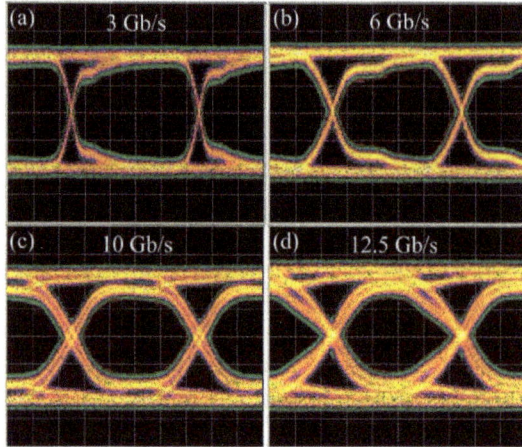

Figure 9. Measured eye diagrams of the total microwave photonics link at (**a**) 3 Gb/s, (**b**) 6 Gb/s, (**c**) 10 Gb/s, and (**d**) 12.5 Gb/s.

5. Conclusions

We demonstrated a flexible thermo-optic variable attenuator based on LRSPP waveguide that can be used to transmit and equalize high frequency microwave signals in microwave photonic systems. The controllable optical attenuation has been realized by applied different driving voltage to the thermo-optic VOA. The experimental results show that the maximum optical attenuation of the VOA is over 28 dB at a driving voltage of 4.17 V. The maximum attenuation of microwave signals is about 55 dB. The flexible thermo-optic VOA using LRSPP waveguide is promising for chip-scale interconnection in high-density photonic integrated circuits and data transmission, amplitude control, and equalization in microwave photonic systems.

Author Contributions: J.T. and T.Z. conceived and designed the experiments; Y.-R.L. and L.-J.Z. performed the experiments; X.-C.F. and X.-M.X. analyzed the data; G.Q. and N.Z. contributed reagents and materials; J.T. wrote the paper.

Funding: This work is supported by Ministry of Science and Technology (MOST) under Grant Number 2017YFA0205800, National Natural Science Foundation of China (NSFC) under grant numbers 11734005 and the Fundamental Research Funds for the Central Universities under grant numbers 2242018k1G020 and 2242015KD006.

Acknowledgments: We appreciate Sami Iqbal for the manuscript revision. We also appreciate Wei Li at School of Information Science and Engineering, Southeast University for granting us access to their BER test system.

Conflicts of Interest: The authors declare no conflict of interest.

References

1. Rogers, J.A.; Someya, T.; Huang, Y. Materials and mechanics for stretchable electronics. *Science* **2010**, *327*, 1603–1607. [CrossRef] [PubMed]
2. Khan, Y.; Ostfeld, A.E.; Lochner, C.M.; Pierre, A.; Arias, A.C. Monitoring of vital signs with flexible and wearable medical devices. *Adv. Mater.* **2016**, *28*, 4373–4395. [CrossRef] [PubMed]

3. Fan, F.R.; Tang, W.; Wang, Z.L. Flexible nanogenerators for energy harvesting and self-powered electronics. *Adv. Mater.* **2016**, *28*, 4283–4305. [CrossRef] [PubMed]
4. Kim, J.T.; Ju, J.J.; Park, S.; Kim, M.S.; Park, S.K.; Lee, M.H. Chip-to-chip optical interconnect using gold long-range surface plasmon polariton waveguides. *Opt. Express* **2008**, *16*, 13133–13138. [CrossRef] [PubMed]
5. Prajzler, V.; Neruda, M.; Nekvindova, P. Flexible multimode polydimethyl-diphenylsiloxane optical planar waveguides. *J. Mater. Sci. Mater. Electron.* **2018**, *29*, 5878–5884. [CrossRef]
6. Tsuchizawa, T.; Yamada, K.; Watanabe, T.; Park, S.; Nishi, H.; Kou, R.; Shinojima, H.; Itabashi, S.-I. Monolithic integration of silicon-, germanium-, and silica-based optical devices for telecommunications applications. *IEEE J. Sel. Top. Quantum Electron.* **2011**, *17*, 516–525. [CrossRef]
7. Dong, P. Silicon photonic integrated circuits for wavelength-division multiplexing applications. *IEEE J. Sel. Top. Quantum Electron.* **2016**, *22*, 370–378. [CrossRef]
8. Capmany, J.; Mora, J.; Ortega, B.; Pastor, D. Microwave photonic filters using low-cost sources featuring tunability, reconfigurability and negative coefficients. *Opt. Express* **2005**, *13*, 1412–1417. [CrossRef] [PubMed]
9. Li, X.; Dong, J.; Yu, Y.; Zhang, X. A tunable microwave photonic filter based on an all-optical differentiator. *IEEE Photonics Technol. Lett.* **2011**, *23*, 308–310. [CrossRef]
10. Marpaung, D.; Roeloffzen, C.; Heideman, R.; Leinse, A.; Sales, S.; Capmany, J. Integrated microwave photonics. *Laser Photonics Rev.* **2013**, *7*, 506–538. [CrossRef]
11. Wu, J.Y.; Xu, X.Y.; Nguyen, T.G.; Chu, S.T.; Little, B.E.; Morandotti, R.; Mitchell, A.; Moss, D.J. Rf photonics: An optical microcombs' perspective. *IEEE J. Sel. Top. Quantum Electron.* **2018**, *24*, 1–20. [CrossRef]
12. Wang, L.F.; Song, Q.Q.; Wu, J.Y.; Chen, K.X. Low-power variable optical attenuator based on a hybrid sion-polymer s-bend waveguide. *Appl. Opt.* **2016**, *55*, 969–973. [CrossRef] [PubMed]
13. Maese-Novo, A.; Zhang, Z.Y.; Irmscher, G.; Polatynski, A.; Mueller, T.; de Felipe, D.; Kleinert, M.; Brinker, W.; Zawadzki, C.; Keil, N. Thermally optimized variable optical attenuators on a polymer platform. *Appl. Opt.* **2015**, *54*, 569–575. [CrossRef]
14. Chen, S.T.; Shi, Y.C.; He, S.L.; Dai, D.X. Variable optical attenuator based on a reflective mach-zehnder interferometer. *Opt. Commun.* **2016**, *361*, 55–58. [CrossRef]
15. Mohsin, M.; Schall, D.; Otto, M.; Chmielak, B.; Porschatis, C.; Bolten, J.; Neumaier, D. Graphene based on-chip variable optical attenuator operating at 855 nm wavelength. *Opt. Express* **2017**, *25*, 31660–31669. [CrossRef] [PubMed]
16. Yu, Y.Y.; Sun, X.Q.; Ji, L.T.; He, G.B.; Wang, X.B.; Yi, Y.J.; Chen, C.M.; Wang, F.; Zhang, D.M. The 650-nm variable optical attenuator based on polymer/silica hybrid waveguide. *Chin. Phys. B* **2016**, *25*, 054101. [CrossRef]
17. Sun, X.; Xie, Y.; Liu, T.; Chen, C.; Wang, F.; Zhang, D. Variable optical attenuator based on long-range surface plasmon polariton multimode interference coupler. *J. Nanomater.* **2014**, *2014*, 1. [CrossRef]
18. Gu, Y.L.; Chen, C.M.; Zheng, Y.; Shi, Z.S.; Wang, X.B.; Yi, Y.J.; Jiang, T.C.; Sun, X.Q.; Wang, F.; Cui, Z.C.; et al. Heat-induced multimode interference variable optical attenuator based on novel organic-inorganic hybrid materials. *J. Opt.* **2015**, *17*, 085802. [CrossRef]
19. Oh, M.-C.; Chu, W.-S.; Shin, J.-S.; Kim, J.-W.; Kim, K.-J.; Seo, J.-K.; Lee, H.-K.; Noh, Y.-O.; Lee, H.-J. Polymeric optical waveguide devices exploiting special properties of polymer materials. *Opt. Commun.* **2016**, *362*, 3–12. [CrossRef]
20. Noh, Y.O.; Lee, C.H.; Kim, J.M.; Hwang, W.Y.; Won, Y.H.; Lee, H.J.; Han, S.G.; Oh, M.C. Polymer waveguide variable optical attenuator and its reliability. *Opt. Commun.* **2004**, *242*, 533–540. [CrossRef]
21. Sorger, V.J.; Oulton, R.F.; Ma, R.M.; Zhang, X. Toward integrated plasmonic circuits. *MRS Bull.* **2012**, *37*, 728–738. [CrossRef]
22. Fang, Y.; Sun, M. Nanoplasmonic waveguides: Towards applications in integrated nanophotonic circuits. *Light Sci. Appl.* **2015**, *4*, e294. [CrossRef]
23. Zhang, X.Y.; Hu, A.; Wen, J.Z.; Zhang, T.; Xue, X.J.; Zhou, Y.; Duley, W.W. Numerical analysis of deep sub-wavelength integrated plasmonic devices based on semiconductor-insulator-metal strip waveguides. *Opt. Express* **2010**, *18*, 18945–18959. [CrossRef] [PubMed]
24. Gramotnev, D.K.; Bozhevolnyi, S.I. Plasmonics beyond the diffraction limit. *Nat. Photonics* **2010**, *4*, 83–91. [CrossRef]
25. Berini, P. Long-range surface plasmon polaritons. *Adv. Opt. Photonics* **2009**, *1*, 484–588. [CrossRef]

26. Dabos, G.; Manolis, A.; Papaioannou, S.; Tsiokos, D.; Markey, L.; Weeber, J.C.; Dereux, A.; Giesecke, A.L.; Porschatis, C.; Chmielak, B.; et al. Cmos plasmonics in wdm data transmission: 200 Gb/s (8 × 25 Gb/s) transmission over aluminum plasmonic waveguides. *Opt. Express* **2018**, *26*, 12469–12478. [CrossRef] [PubMed]

27. Dabos, G.; Ketzaki, D.; Manolis, A.; Markey, L.; Weeber, J.C.; Dereux, A.; Giesecke, A.L.; Porschatis, C.; Chmielak, B.; Tsiokos, D.; et al. Plasmonic stripes in aqueous environment co-integrated with si3n4 photonics. *IEEE Photonics J.* **2018**, *10*, 1–8. [CrossRef]

28. Gagnon, G.; Lahoud, N.; Mattiussi, G.A.; Berini, P. Thermally activated variable attenuation of long-range surface plasmon-polariton waves. *J. Lightwave Technol.* **2006**, *24*, 4391–4402. [CrossRef]

29. Lee, J.; Lu, F.; Belkin, M.A. Broadly wavelength tunable bandpass filters based on long-range surface plasmon polaritons. *Opt. Lett.* **2011**, *36*, 3744–3746. [CrossRef] [PubMed]

30. Papaioannou, S.; Kalavrouziotis, D.; Vyrsokinos, K.; Weeber, J.-C.; Hassan, K.; Markey, L.; Dereux, A.; Kumar, A.; Bozhevolnyi, S.I.; Baus, M.; et al. Active plasmonics in wdm traffic switching applications. *Sci. Rep.* **2012**, *2*, 652. [CrossRef] [PubMed]

31. Kim, J.T.; Chung, K.H.; Choi, C.-G. Thermo-optic mode extinction modulator based on graphene plasmonic waveguide. *Opt. Express* **2013**, *21*, 15280–15286. [CrossRef] [PubMed]

32. Haffner, C.; Heni, W.; Fedoryshyn, Y.; Niegemann, J.; Melikyan, A.; Elder, D.L.; Baeuerle, B.; Salamin, Y.; Josten, A.; Koch, U.; et al. All-plasmonic mach-zehnder modulator enabling optical high-speed communication at the microscale. *Nat. Photonics* **2015**, *9*, 525–529. [CrossRef]

33. Fan, H.; Berini, P. Thermo-optic characterization of long-range surface-plasmon devices in cytop. *Appl. Opt.* **2013**, *52*, 162–170. [CrossRef] [PubMed]

34. Zografopoulos, D.C.; Beccherelli, R. Plasmonic variable optical attenuator based on liquid-crystal tunable stripe waveguides. *Plasmonics* **2013**, *8*, 599–604. [CrossRef]

35. Leosson, K.; Rosenzveig, T.; Hermannsson, P.G.; Boltasseva, A. Compact plasmonic variable optical attenuator. *Opt. Express* **2008**, *16*, 15546–15552. [CrossRef] [PubMed]

36. Nikolajsen, T.; Leosson, K.; Bozhevolnyi, S.I. In-line extinction modulator based on long-range surface plasmon polaritons. *Opt. Commun.* **2005**, *244*, 455–459. [CrossRef]

37. Vernoux, C.; Chen, Y.T.; Markey, L.; Sparchez, C.; Arocas, J.; Felder, T.; Neitz, M.; Brusberg, L.; Weeber, J.C.; Bozhevolnyi, S.I.; et al. Flexible long-range surface plasmon polariton single-mode waveguide for optical interconnects. *Opt. Mater. Express* **2018**, *8*, 469–484. [CrossRef]

38. Nikolajsen, T.; Leosson, K.; Salakhutdinov, I.; Bozhevolnyi, S.I. Polymer-based surface-plasmon-polariton stripe waveguides at telecommunication wavelengths. *Appl. Phys. Lett.* **2003**, *82*, 668–670. [CrossRef]

39. Wu, X.P.; Liu, W.; Yuan, Z.L.; Liang, X.R.; Chen, H.; Xu, X.H.; Tang, F.F. Low power consumption voa array with air trenches and curved waveguide. *IEEE Photonics J.* **2018**, *10*, 1–8. [CrossRef]

40. Sun, S.Q.; Niu, D.H.; Sun, Y.; Wang, X.B.; Yang, M.; Yi, Y.J.; Sun, X.Q.; Wang, F.; Zhang, D.M. Design and fabrication of all-polymer thermo-optic variable optical attenuator with low power consumption. *Appl. Phys. Mater. Sci. Process.* **2017**, *123*, 646. [CrossRef]

41. Ren, M.Z.; Zhang, J.S.; An, J.M.; Wang, Y.; Wang, L.L.; Li, J.G.; Wu, Y.D.; Yin, X.J.; Hu, X.W. Low power consumption 4-channel variable optical attenuator array based on planar lightwave circuit technique. *Chin. Phys. B* **2017**, *26*, 074221. [CrossRef]

micromachines

MDPI

Article

An Organic Flexible Artificial Bio-Synapses with Long-Term Plasticity for Neuromorphic Computing

Tian-Yu Wang, Zhen-Yu He, Lin Chen * ⓘ, Hao Zhu, Qing-Qing Sun, Shi-Jin Ding, Peng Zhou and David Wei Zhang

State Key Laboratory of ASIC and System, School of Microelectronics, Fudan University, Shanghai 200433, China; wangtianyu16@fudan.edu.cn (T.-Y.W.); 17212020012@fudan.edu.cn (Z.-Y.H.); hao_zhu@fudan.edu.cn (H.Z.); qqsun@fudan.edu.cn (Q.-Q.S.); sjding@fudan.edu.cn (S.-J.D.); pengzhou@fudan.edu.cn (P.Z.); dwzhang@fudan.edu.cn (D.W.Z.)
* Correspondence: linchen@fudan.edu.cn; Tel.: +86-135-2402-6812

Received: 19 April 2018; Accepted: 11 May 2018; Published: 15 May 2018

Abstract: Artificial synapses, with synaptic plasticity, are the key components of constructing the neuromorphic computing system and mimicking the bio-synaptic function. Traditional synaptic devices are based on silicon and inorganic materials, while organic electronics can open up new opportunities for flexible devices. Here, a flexible artificial synaptic device with an organic functional layer was proposed. The organic device showed good switching behaviors such as ON/OFF ratio over 100 at low operation voltages. The set and reset voltages were lower than 0.5 V and -0.25 V, respectively. The long-term plasticity, spike-timing-dependent plasticity learning rules (STDP), and forgetting function were emulated using the device. The retention times of the excitatory and inhibitory post-synaptic currents were both longer than 60 s. The long-term plasticity was repeatable without noticeable degradation after the application of five voltage pulse cycles to the top electrode. These results indicate that our organic flexible device has the potential to be applied in bio-inspired neuromorphic systems.

Keywords: flexible organic electronics; artificial synapses; neuromorphic computing; long-term plasticity

1. Introduction

The human brain can be seen as an effective system that is capable of analyzing complicated tasks through the integration of storage and computation [1,2]. To date, a neuromorphic computing system has been proposed and developed to overcome the bottleneck of classical von Neumann computers [3]. It has been widely recognized that fabricating an artificial electronic device with the function of mimicking the behaviors of a bio-synapse is necessary to realize neuromorphic computing. Many synaptic behaviors have been emulated using artificial synaptic devices, including long-term potentiation (LTP), long-term depression (LTD), paired-pulse facilitation (PPF), and STDP (spike-timing-dependent plasticity) [4]. The conductance of devices should be modulated gradually and simulate weight changes of bio-synapses [5,6], which are the fundamental to achieving synaptic plasticity. In recent years, various devices including CMOS transistors, resistive random access memory (RRAM), ferroelectric random access memory (FeRAM), and phase-change memory (PCM) have been demonstrated exhibiting such synaptic behaviors [4,6–8]. Among them, RRAM with the advantages of high-integration, low-power consumption, and simple structure, has become one of the promising candidates for the applications in neuromorphic computing.

On the other hand, flexible electronics has attracted more interests of researchers and has been widely studied because of the potential in future wearable devices [9]. Flexible electronics are more portable and deformable in comparison with silicon-based devices [10–12]. However, most RRAM-based synaptic memories are composed of various inorganic materials, such as HfO_x, Al_2O_3, and ZnO [13–15]. These inorganic materials usually require high temperature treatment steps with poor stretchability. In addition, the intrinsic properties of these materials are not compatible with a flexible substrate, which cannot meet the development and applications of flexible electronics [15–17]. Therefore, it is urgent to find a type of material suitable for flexible devices. There are many reports showing that organic materials can avoid high temperature treatment [18,19], which can be applied as the functional layers of flexible RRAM. Organic polymers have the advantages of simple preparation process at room temperature, low cost, and good stretchability. Poly(3,4-ethylenedioxythiophene): poly(styrenesulfonate) (PEDOT:PSS) is one of the common polymers with excellent stretchability [20,21], which has been proven to have resistive switching behaviors [22] and synaptic plasticity separately on the rigid substrate [23,24]. However, the realization of abrupt bipolar resistive switching characteristics and mimicking synaptic behaviors at the same time on a flexible substrate with PEDOT:PSS has not been reported.

Here, we fabricated a flexible RRAM based on PEDOT:PSS and examined its current response to different voltages. The device showed excellent resistive switching characteristics under direct-current sweep. It turned from high resistance state (HRS) to low resistance state (LRS) and came back to HRS at low operation voltages. Furthermore, good synaptic plasticity, including LTP, LTD, forgetting curve, and STDP were demonstrated under pulse chains in this flexible synaptic device. The controllable conductance is related to the transformation and migration of $PEDOT^+$ [25]. These results demonstrate the feasibility of flexible PEDOT:PSS-based RRAM used as artificial synapses for neuromorphic computing and the potential for wearable electronics applications [26,27].

2. Materials and Methods

The preparation of active layer was processed with the solution of PEDOT:PSS (Clevios PH1000), which was purchased from Heraeus (Germany). Before coating, the PEDOT:PSS solution was filtered through a micro filter membrane with pore size of 0.22 μm.

The synaptic device made by us has a structure of Indium Tin Oxides (ITO)/PEDOT:PSS/Au with a cross-sectional junction circle of 200 um diameter, as shown in Figure 1a. Polyethylene terephthalate (PET) was adopted as the flexible substrate. The substrate of ITO-coated PET was cleaned by mixing detergent of acetone and isopropyl alcohol (IPA) in an ultrasonic bath for 5 min. Then, the substrate was treated with oxygen plasma at 150 W for 3 min, and the film of PEDOT:PSS was spin-coated on the ITO electrode followed by baking at 120 °C for 10 min on a hotplate. The electrode of Au was deposited on PEDOT:PSS with a shadow mask by physical vapor deposition(PVD, Sputter system M362, SPECS), as shown in Figure 1b.

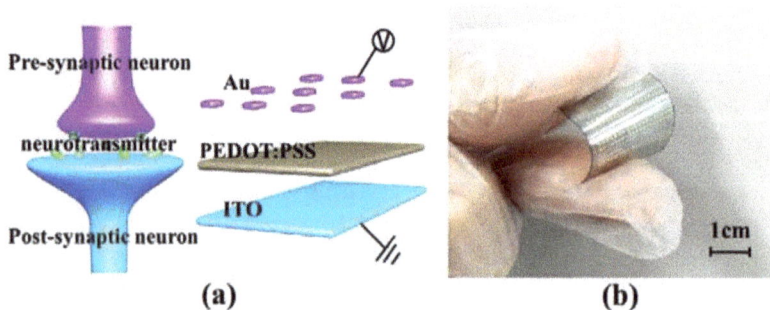

Figure 1. (**a**) The schematic structure of a bio-synapse and the corresponded PEDOT:PSS-based RRAM; (**b**) optical image of our flexible synaptic device in the bend state.

All measurements of PEDOT:PSS-based RRAM were at room temperature, and atmospheric pressure biased the top electrode (Au) and grounded the bottom electrode (ITO). The electrical characteristics of the device were performed using Agilent B1500A and B1525 semiconductor parameter analyzer. The resistive switching characteristics were achieved by applying DC voltage with B1500A to the top electrode of the device. Two pulse channels of the B1525 were used to input pre- and post-synaptic pulses and induce synaptic behaviors of our device. The surface topography of PEDOT:PSS film was obtained by field emission scanning electron microscope (FESEM, ZEISS-SIGMA HD), as shown in Figure S1.

3. Results and Discussion

The two terminal structure of the flexible PEDOT:PSS-based RRAM is suitable for large-scale preparation of artificial synaptic arrays [28]. As shown in Figure 1a, the top electrode and bottom electrode corresponds to pre- and post-synaptic neuron, respectively.

Figure 2 demonstrated the resistive switching characteristics of the flexible RRAM. The forming voltage was ~3 V, and the operation voltage was very low, i.e., the set voltage is lower than 0.5 V, and the reset voltage is lower than −0.25 V. A 0 V→1 V→0 V→−0.8 V→0 V DC voltage cycle was applied to the top electrode of Au with the ITO bottom electrode grounded. We realized the typical bipolar characteristics for the ITO/PEDOT:PSS/Au. The current-voltage (I–V) characteristics of our device through 15 cycles were measured. The device showed stable set operations under a compliance current (CC) of 100 uA in Figure 2a. The device showed high ON/OFF ratio larger than 100, as shown in Figure 2b. The resistive switching behaviors were linked to the regeneration and rupture of PEDOT$^+$ conductive paths because of the injection and extraction of the hole in the film under the positive and negative voltages [29]. When the positive voltage was applied to the top electrode, the hole injected to PEDOT^0, and it turned to PEDOT$^+$. PEDOT$^+$ was more conductive than PEDOT^0 and could be accumulated to form a conductive path in the film.

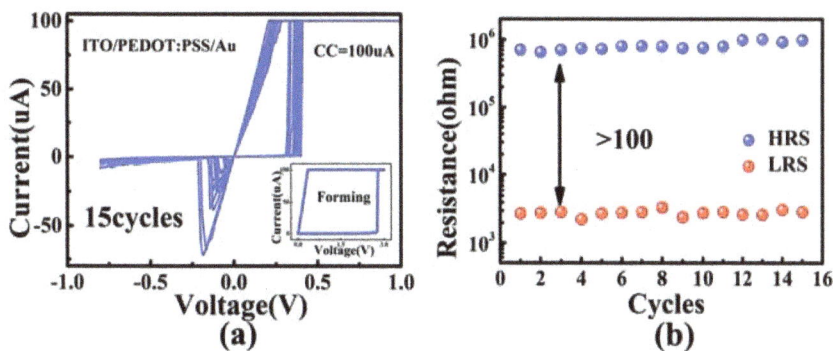

Figure 2. (a) The I-V curves of the flexible PEDOT:PSS-based RRAM. Inset shows the forming process of the device; (b) the HRS and LRS measured by DC sweeping of the device.

To examine the application of artificial synaptic device in synaptic plasticity, we applied continuous positive or negative voltages to the top electrode of Au. Both positive and negative voltages were carefully chosen to avoid abrupt resistance switching behaviors. As shown in Figure 3a, there were five consecutive sweeps of positive voltages or negative voltages. With the positive voltages swept from 0 V→3 V→0 V, the conductance decreased after each cycle (Figure 2a). In contrast, the conductance increased gradually under 5 negative DC sweeps (0 V→−2 V→0 V), which was similar as potential behaviors in bio-synapses. To clearly show the successful modulation of synaptic weights in our device, pulse training mode was utilized. The pulse amplitude was 2 V, and the pulse width was 10 ms without intervals, which was indicated by the blue lines in Figure 3b. The red lines were the current response according to 10 consecutive pulse training. The conductance of our device was potentiated by 5 negative bias pulses and depressed by 5 positive bias pulses, showing the potential for LTP and LTD under pulse tests.

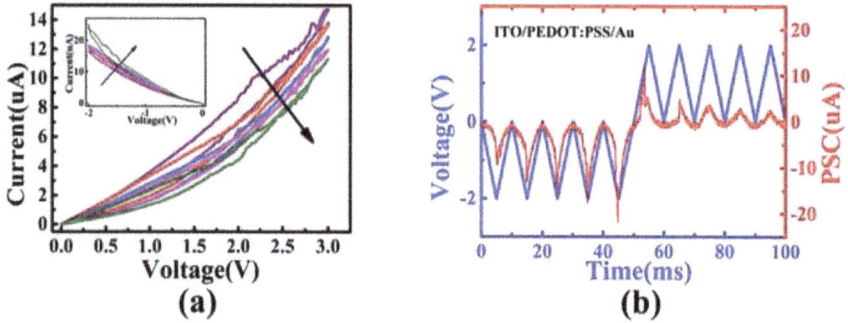

Figure 3. (a) I-V curves during five DC sweeps from 0 V to 3 V of PEDOT:PSS-based RRAM. Inset shows the I-V curves under five negative voltage sweeps from 0 V to −2 V; (b) the modulated currents (red lines) under 5 constructive negative pulse (−2 V, 10 ms) and positive pulse (2 V, 10 ms) trains (blue lines).

Gradual modulated conductance is important for a synaptic device towards neuromorphic computing [30]. We applied 300 negative pulses (−1.5 V, 10 ms) followed by 300 read operations with pulses of 0.1 V. The post-synaptic current under the read voltages increased gradually, which corresponded to the LTP behavior of bio-synapses. Similarly, 300 positive pulses (1 V, 10 ms) and read pulses of 0.1 V were applied to our devices, which successfully emulated the LTD behavior (Figure 4a). The response of current during measurement of LTD was shown in Figure S2, and the schematic of pulse waveform was shown in Figure S3. Furthermore, we repeated the pulse chains 5 times, and the device showed excellent endurance (Figure 4b). The device showed STDP learning rules in Figure 4c, which described the relationship between weight change (ΔW) and time interval (Δt). The time interval was defined as follows:

$$\Delta t = t_{post} - t_{pre}, \tag{1}$$

in which the t_{post} and t_{pre} were the time of pulse come to the post-synaptic and pre-synaptic electrode. A pair of pulses was applied to the pre-synaptic electrode (−1.5 V, 10 ms) and post-synaptic electrode (1.5 V, 10 ms). Additionally, the weight change was described as follows:

$$\Delta W = (G_t - G_0)/G_0 \tag{2}$$

in which G_t was the conductance of the device at the time node of "t" and G_0 was the initial conductance of the device at the time node of "$t = 0$". When the pulse was exerted earlier on the pre-synapse, Δt was greater than 0. The weights increased, which indicated that the connection relationship between two synapses was potentiated. In contrast, the weights decreased, which indicated that the connection strength of two synapses was weakened ($\Delta t < 0$).

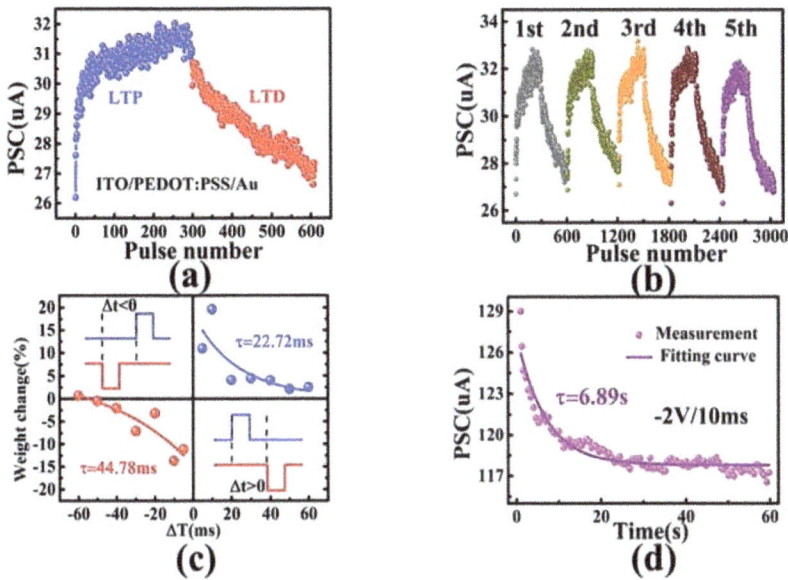

Figure 4. (**a**) The LTP and LTD under 300 negative pulses (−1.5 V, 10 ms) and 300 positive pulses (1 V, 10 ms) of our PEDOT:PSS-based RRAM; (**b**) Five operations of LTP and LTD; (**c**) simulation of STDP by changing the pulse intervals of pre- and post-synaptic spiking; (**d**) forgetting curve after a single pulse (−2 V,10 ms). The responded current was read at 0.1 V.

Forgetting curve is a common bio-synaptic behavior, which has been studied widely and reported [2,11,31]. The relaxation of post-synaptic current after spiking pulse to the pre-synaptic electrode could be used for mimicking the forgetting function in psychology. Figure S4 showed the whole process during the measurement of the forgetting function of the device. As shown in Figure 4d, after applying one pulse (−2 V, 10 ms) to the pre-synaptic electrode of Au, currents of post-synaptic electrode were recorded at the read voltage of 0.1 V for 60 s. The relationship between post-synaptic current and time was fitted by an exponential decay equation as follows:

$$I_t = I_0 + A \exp(-t/\tau), \tag{3}$$

in which I_t and I_0 were the memory current at time of t and stabilized state, A was the prefactor, and τ was the relaxation time constant, which illustrated the forgetting speed of the memory. The forgetting curve was fitted based on Equation (3), in which τ was 6.89 s.

As a nonvolatile memory, retention characteristic is a significant indicator. We emulated the inhibitory and excitatory features of bio-synapses using post-synaptic current. Upon the application of positive (1 V, 10 ms) and negative (−1.5 V, 10 ms) pulses, the post-synaptic current could hold at least 60 s (Figure 5). Excitatory or inhibitory of PSC could modulate the synaptic connection synergistically. After a negative pulse (−1.5 V, 10 ms) was implemented at the top electrode, the PSC increased obviously, and the change of current between the resting current after 60 s and the instantly stimulated current was kept above 60%. After a positive pulse (1 V, 10 ms) was implemented at the top electrode, the PSC decreased obviously, and the change of current was kept above 100% after 60 s. The life time of the PSC was much longer than 60 s, indicating that LTP and LTD could be realized by multiple pulses with time intervals shorter than 60 s.

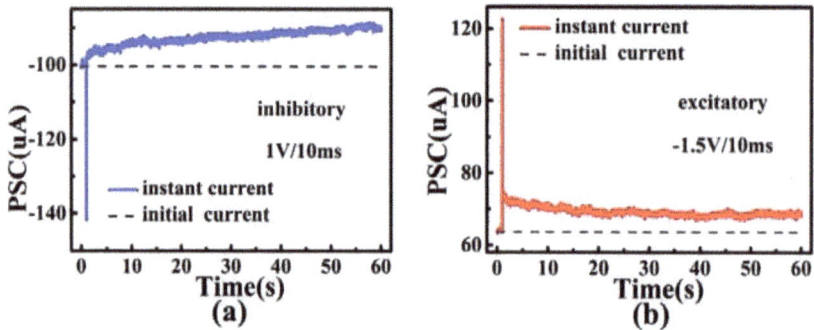

Figure 5. The retention behaviors of (**a**) inhibitory of post-synaptic currents under one pulse (1 V, 10 ms) and (**b**) excitatory features of post-synaptic currents under one pulse (−1.5 V,10 ms).

4. Conclusions

We fabricated a flexible PEDOT:PSS-based artificial synaptic device that exhibited not only abrupt resistive switching of binary characteristic but also gradual muti-level conductance modulation used for mimicking synaptic plasticity. The Au top electrode and ITO bottom electrode corresponded to the pre- and post-synapse. We applied voltage to the top electrode and recorded the responded current of the bottom electrode to assess the device characteristics. The device showed good resistive switching behaviors with the ON/OFF ratio larger than 100. The set and reset voltage was lower than 0.5 V and −0.25 V, respectively. Besides, LTP, LTD, STDP, and forgetting function were manifested in our device. After five repeatable tests with 300 positive and 300 negative pulses for LTD and LTP, there was no obvious degradation observed in the device. These results suggested that our two terminal organic flexible RRAMs had the potential for neuromorphic computing.

Supplementary Materials: The following are available online at http://www.mdpi.com/2072-666X/9/5/239/s1: Figure S1: SEM image of PEDOT:PSS film; Figure S2: transient response during measurement of LTD; Figure S3: continuous pulses used for measuring the LTD; and Figure S4: the response of currents during measuring the forgetting curve.

Author Contributions: L.C., Q.-Q.S., and D.W.Z. conceived and designed the experiments; T.-Y.W. performed the experiments; T.-Y.W., Z.-Y.H. and H.Z. analyzed the data; S.-J.D. and P.Z. contributed some of materials and analysis tools; T.-Y.W., L.C., and Z.-Y.H. co-wrote the paper.

Acknowledgments: The work was supported by the National Nature Science Foundation of China (61704030, 61376092, and 61427901), 02 State Key Project (2017ZX02315005), Shanghai Rising-Star Program (14QA1400200), Shanghai Educational Development Foundation, Program of Shanghai Subject Chief Scientist (14XD1400900), the S&T Committee of Shanghai (14521103000, 15DZ1100702, 15DZ1100503), and "Chen Guang" project supported by Shanghai Municipal Education Commission and Shanghai Education Development Foundation.

Conflicts of Interest: The authors declare no conflict of interest.

References

1. Jeong, D.S.; Kim, K.M.; Kim, S.; Choi, B.J.; Hwang, C.S. Neuromorphic Computing: Memristors for Energy—Efficient New Computing Paradigms (Adv. Electron. Mater. 9/2016). *Adv. Electron. Mater.* **2016**, 2. [CrossRef]
2. Yan, X.; Zhao, J.; Liu, S.; Zhou, Z.; Liu, Q.; Chen, J.; Liu, X.Y. Memristor with Ag-Cluster-Doped TiO$_2$ Films as Artificial Synapse for Neuroinspired Computing. *Adv. Funct. Mater.* **2018**, *28*, 1705320. [CrossRef]
3. Strukov, D.B. Nanotechnology: Smart connections. *Nature* **2011**, *476*, 403–405. [CrossRef] [PubMed]
4. Milo, V.; Ielmini, D.; Chicca, E. In Attractor Networks and Associative Memories with STDP Learning in RRAM Synapses. In *Proceedings of the IEEE International Electron Devices Meeting, San Francisco, CA, USA, 2–6 December 2017*; IEEE: Piscataway, NJ, USA, 2017; pp. 11–12.

5. Kim, M.K.; Lee, J.S. Short-Term Plasticity and Long-Term Potentiation in Artificial Biosynapses with Diffusive Dynamics. *ACS Nano* **2018**, *12*, 1680–1687. [CrossRef] [PubMed]
6. Boyn, S.; Grollier, J.; Lecerf, G.; Xu, B.; Locatelli, N.; Fusil, S.; Girod, S.; Carrétéro, C.; Garcia, K.; Xavier, S. Learning through ferroelectric domain dynamics in solid-state synapses. *Nat. Commun.* **2017**, *8*, 14736. [CrossRef] [PubMed]
7. Ambrogio, S.; Balatti, S.; Milo, V.; Carboni, R.; Wang, Z.Q.; Calderoni, A.; Ramaswamy, N.; Ielmini, D. Neuromorphic Learning and Recognition with One-Transistor-One-Resistor Synapses and Bistable Metal Oxide RRAM. *IEEE Trans. Electron. Dev.* **2016**, *63*, 1508–1515. [CrossRef]
8. Ren, K.; Li, R.; Chen, X.; Wang, Y.; Shen, J.; Xia, M.; Lv, S.; Ji, Z.; Song, Z. Controllable SET process in O-Ti-Sb-Te based phase change memory for synaptic application. *Appl. Phys. Lett.* **2018**, *112*, 73106. [CrossRef]
9. Park, J.; Kim, J.; Kim, S.Y.; Cheong, W.H.; Jang, J.; Park, Y.G.; Na, K.; Kim, Y.T.; Heo, J.H.; Chang, Y.L. Soft, smart contact lenses with integrations of wireless circuits, glucose sensors, and displays. *Sci. Adv.* **2018**, *4*, 9841. [CrossRef] [PubMed]
10. Shang, J.; Xue, W.; Ji, Z.; Liu, G.; Niu, X.; Yi, X.; Pan, L.; Zhan, Q.; Xu, X.H.; Li, R.W. Highly flexible resistive switching memory based on amorphous-nanocrystalline hafnium oxide films. *Nanoscale* **2017**, *9*, 7037–7046. [CrossRef] [PubMed]
11. Park, Y.; Lee, J.S. Artificial Synapses with Short- and Long-Term Memory for Spiking Neural Networks Based on Renewable Materials. *ACS Nano* **2017**, *11*, 8962–8969. [CrossRef] [PubMed]
12. Khiat, A.; Cortese, S.; Serb, A.; Prodromakis, T. Resistive switching of Pt/TiOx/Pt devices fabricated on flexible Parylene-C substrates. *Nanotechnology* **2017**, *28*, 25303. [CrossRef] [PubMed]
13. Brivio, S.; Frascaroli, J.; Spiga, S. Role of Al doping in the filament disruption in HfO$_2$ resistance switches. *Nanotechnology* **2017**, *28*, 395202. [CrossRef] [PubMed]
14. Simanjuntak, F.; Chandrasekaran, S.; Pattanayak, B.; Lin, C.C.; Tseng, T.Y. Peroxide Induced Volatile and Non-volatile Switching Behavior in ZnO-based Electrochemical Metallization Memory Cell. *Nanotechnology* **2017**, *28*. [CrossRef] [PubMed]
15. Lin, Y.D.; Chen, P.S.; Lee, H.Y.; Chen, Y.S.; Rahaman, S.Z.; Tsai, K.H.; Hsu, C.H.; Chen, W.S.; Wang, P.H.; King, Y.C. Retention Model of TaO/HfOx and TaO/AlOx RRAM with Self-Rectifying Switch Characteristics. *Nanoscale Res. Lett* **2017**, *12*, 407. [CrossRef] [PubMed]
16. Zhang, K.; Sun, K.; Wang, F.; Han, Y.; Jiang, Z.; Zhao, J.; Wang, B.; Zhang, H.; Jian, X.; Wong, H.S.P. Ultra-Low Power Ni/HfO$_2$/TiO$_x$/TiN Resistive Random Access Memory With Sub-30-nA Reset Current. *IEEE Electr. Dev. Lett.* **2015**, *36*, 1018–1020. [CrossRef]
17. Chen, Z.; Zhang, F.; Chen, B.; Zheng, Y.; Gao, B.; Liu, L.; Liu, X.; Kang, J. High-performance HfOx/AlOy—Based resistive switching memory cross-point array fabricated by atomic layer deposition. *Nanoscale Res. Lett.* **2015**, *10*, 70. [CrossRef] [PubMed]
18. Kong, L.A.; Sun, J.; Qian, C.; Fu, Y.; Wang, J.; Yang, J.; Gao, Y. Long-term synaptic plasticity simulated in ionic liquid/polymer hybrid electrolyte gated organic transistors. *ORG Electron.* **2017**, *47*, 126–132. [CrossRef]
19. Van, D.B.Y.; Lubberman, E.; Fuller, E.J.; Keene, S.T.; Faria, G.C.; Agarwal, S.; Marinella, M.J.; Alec, T.A.; Salleo, A. A non-volatile organic electrochemical device as a low-voltage artificial synapse for neuromorphic computing. *Nat. Mater.* **2017**, *16*, 414–418.
20. Wang, Y.; Zhu, C.; Pfattner, R.; Yan, H.; Jin, L.; Chen, S.; Molinalopez, F.; Lissel, F.; Liu, J.; Rabiah, N.I. A highly stretchable, transparent, and conductive polymer. *Sci. Adv.* **2017**, *3*, e1602076. [CrossRef] [PubMed]
21. Yu, J.C.; Jang, J.I.; Bo, R.L.; Lee, G.W.; Han, J.T.; Song, M.H. Highly Efficient Polymer-Based Optoelectronic Devices Using PEDOT:PSS and a GO Composite Layer as a Hole Transport Layer. *ACS Appl. Mater. Interfaces* **2014**, *6*, 2067–2073. [CrossRef] [PubMed]
22. Wang, Z.; Zeng, F.; Yang, J.; Chen, C.; Pan, F. Resistive switching induced by metallic filaments formation through poly(3,4-ethylene-dioxythiophene): poly(styrenesulfonate). *ACS Appl. Mater. Interfaces* **2012**, *4*, 447–453. [CrossRef] [PubMed]
23. Zeng, F.; Li, S.; Yang, J.; Pan, F.; Guo, D. Learning processes modulated by the interface effects in a Ti/conducting polymer/Ti resistive switching cell. *RSC Adv.* **2014**, *4*, 14822–14828. [CrossRef]
24. Choi, H.Y.; Wu, C.; Chang, H.B.; Kim, T.W. Organic electronic synapses with pinched hystereses based on graphene quantum-dot nanocomposites. *NPG Asia Mater.* **2017**, *9*, e413. [CrossRef]

25. Chia, P.; Chua, L.; Sivaramakrishnan, S.; Zhuo, J.; Zhao, L.; Sim, W.; Yeo, Y.; Ho, P.K. Injection-induced De-doping in a Conducting Polymer during Device Operation: Asymmetry in the Hole Injection and Extraction Rates. *Adv. Mater.* **2007**, *19*, 4202–4207. [CrossRef]
26. Lipomi, D.J.; Bao, Z. Stretchable and ultraflexible organic electronics. *MRS Bull.* **2017**, *42*, 93–97. [CrossRef]
27. An, B.W.; Shin, J.H.; Kim, S.Y.; Kim, J.; Ji, S.; Park, J.; Lee, Y.; Jang, J.; Park, Y.G.; Cho, E. Smart Sensor Systems for Wearable Electronic Devices. *Polymers (Basel)* **2017**, *9*, 303. [CrossRef]
28. Kim, S.; Du, C.; Sheridan, P.; Ma, W.; Choi, S.; Lu, W.D. Experimental Demonstration of a Second-Order Memristor and Its Ability to Biorealistically Implement Synaptic Plasticity. *Nano Lett.* **2015**, *15*, 2203–2211. [CrossRef] [PubMed]
29. Liu, X.; Ji, Z.; Tu, D.; Shang, L.; Liu, J.; Liu, M.; Xie, C. Organic nonpolar nonvolatile resistive switching in poly(3,4-ethylene-dioxythiophene): Polystyrenesulfonate thin film. *ORG Electron.* **2009**, *10*, 1191–1194. [CrossRef]
30. Yu, S. Neuro-Inspired Computing With Emerging Nonvolatile Memory. *Proc. IEEE* **2018**, *106*, 260–285. [CrossRef]
31. Ohno, T.; Hasegawa, T.; Tsuruoka, T.; Terabe, K.; Gimzewski, J.K.; Aono, M. Short-term plasticity and long-term potentiation mimicked in single inorganic synapses. *Nat. Mater.* **2011**, *10*, 591–595. [CrossRef] [PubMed]

micromachines

MDPI

Article

Stretchable Tattoo-Like Heater with On-Site Temperature Feedback Control

Andrew Stier [1], Eshan Halekote [1], Andrew Mark [1], Shutao Qiao [2], Shixuan Yang [2], Kenneth Diller [3] and Nanshu Lu [1,2,3,4,*]

[1] Department of Electrical and Computer Engineering, University of Texas at Austin, Austin, TX 78712, USA; andrewcstier@gmail.com (A.S.); ehalekote@gmail.com (E.H.); andrewemark@gmail.com (A.M.)

[2] Center for Mechanics of Solids, Structures and Materials, Department of Aerospace Engineering and Engineering Mechanics, University of Texas at Austin, Austin, TX 78712, USA; shutao2011@gmail.com (S.Q.); rock002008@gmail.com (S.Y.)

[3] Department of Biomedical Engineering, University of Texas at Austin, Austin, TX 78712, USA; kdiller@mail.utexas.edu

[4] Texas Materials Institute, the University of Texas at Austin, Austin, TX 78712, USA

* Correspondence: nanshulu@utexas.edu; Tel.: +512-471-4208

Received: 20 February 2018; Accepted: 27 March 2018; Published: 8 April 2018

Abstract: Wearable tissue heaters can play many important roles in the medical field. They may be used for heat therapy, perioperative warming and controlled transdermal drug delivery, among other applications. State-of-the-art heaters are too bulky, rigid, or difficult to control to be able to maintain long-term wearability and safety. Recently, there has been progress in the development of stretchable heaters that may be attached directly to the skin surface, but they often use expensive materials or processes and take significant time to fabricate. Moreover, they lack continuously active, on-site, unobstructive temperature feedback control, which is critical for accommodating the dynamic temperatures required for most medical applications. We have developed, fabricated and tested a cost-effective, large area, ultra-thin and ultra-soft tattoo-like heater that has autonomous proportional-integral-derivative (PID) temperature control. The device comprises a stretchable aluminum heater and a stretchable gold resistance temperature detector (RTD) on a soft medical tape as fabricated using the cost and time effective "cut-and-paste" method. It can be noninvasively laminated onto human skin and can follow skin deformation during flexure without imposing any constraint. We demonstrate the device's ability to maintain a target temperature typical of medical uses over extended durations of time and to accurately adjust to a new set point in process. The cost of the device is low enough to justify disposable use.

Keywords: epidermal electronics; wearable heater; temperature sensor; feedback control

---

1. Introduction

There exists a need for soft, stretchable electronic heating devices that can conform to human skin unobstructively and stay attached during long term use. Such devices can serve a variety of applications in the medical field. As some examples, heat is commonly used in physical therapy following exercise-induced delayed onset muscle soreness (DOMS) [1,2]. Heating injured joints can induce thermal expansion of the collagen tissue and thus reduce pain and stiffness [3–5]. When hypothermia occurs due to anesthesia [6], applying heat to the palms and soles of a patient with distended blood vessels can re-warm the body's core temperature [7,8]. Applying heat over a skin surface can accelerate the transdermal diffusion of chemicals from a drug patch [9,10].

Conventional heaters used to treat muscle pain or joint injuries include electric heat packs [11] and heat wraps [12]. Heat packs do not have very controllable temperature and are heavy and bulky.

Heat wraps are easier to control but are also heavy and their rigidity makes it difficult for them to be worn seamlessly [5]. These products' inability to conform well to skin [13–15] make them less comfortable and also present a more severe problem—lack of uniform and consistent adhesion to the skin surface could lead to air gaps which cause hotspots [16]. These hotspots could burn the skin if the heater is operated near the safety threshold of 43 °C [17]. This can severely limit the range and thus the effectiveness of the conventional heaters.

One heating method that can safely heat the body at temperatures close to 43 °C and the current gold standard for preventing the hypothermia caused by anesthesia, is forced air warming. Forced air warming heats air and pumps it into blankets covering large portions of the patient. While effective at raising the core body temperature, forced air warming has some disadvantages including bulkiness, obstructiveness to surgeries and high cost [18,19].

Recently there has been an expansion of the development of stretchable electronics [20–23]. Methods that have been used to produce these type of electronics include embedding carbon nanotubes (CNTs) in elastomers [24,25], depositing silver (Ag) nanoparticles in polyeruthane [5,26], chemically bonding Ag flakes to CNTs [27,28], combining Ag nanoparticles with elastomeric fibers [29], electrospinning Ag nanofibers onto a flexible substrate [30], constructing CuZr nanotrough networks that function as stretchable electrodes [31], constructing stretchable gold (Au) electrodes from multi-layers of Au nanosheets [32] and patterning metal thin films into serpentine [33,34] or fractal shapes [35,36] to minimize their strain during stretching. This last method has enabled the creation of epidermal electronics—ultrathin, ultrasoft electronics, physiological sensors, and electrical and thermal stimulators—that can adhere and conform to skin surfaces and bend and stretch without breaking, detaching, or imposing any mechanical constraint to the skin [36–38].

With the development of stretchable electronics, stretchable patch heaters have emerged in recent years. Examples include joule heating devices fabricated from soft Ag nanowire composites [5,30,39], copper (Cu) nanowire based fabric [40], stretchable copper zirconium electrodes [31], or stretchable Au serpentines [6,11,29,32]. Using a stretchable and conformable heater could solve the major disadvantages of conventional solid heaters. However, the existing stretchable heaters involve expensive nanomaterials or time-consuming procedures to produce. Moreover, most of them have no method of acquiring temperature feedback from the heater as they are not equipped with any temperature sensors. As a result, most of the reported stretchable heaters do not use temperature feedback to autonomously maintain a set temperature for the heater. One of the exceptions is a wearable fabric heater described by Cheng et al. [40]. This heater is loose enough that it is able to use an unspecified temperature sensor to monitor and control the heater. Of the more tightly conformable heaters, one with temperature feedback only has the functionality of turning the heater off if it gets too hot—aside from that, the temperature feedback is not used to actively control the heat of the heater [9]. The best existing example of a tightly conforming stretchable heater with continuous feedback control is the metal nanofiber heater developed by Jang et al. [30]. This heater uses a thermistor placed on the outer edge of the heater to detect the heater's temperature and uses an unspecified control algorithm to keep the heater at a specified temperature. This set-up relies on the assumption that the temperature at the outer edge of the heater is representative of the temperature of the heater overall but the heat profile across the heater is not actually uniform. Moreover, the thermistor appears to introduce disturbances in the uniformity of the heater.

In other stretchable heaters with temperature sensors, the temperature sensing element is the same as the heating element [9,15]. The biggest disadvantage of this setup is that the dual-purpose element is measuring its own temperature instead of the actual skin temperature. The closest example of a stretchable heater with separate, unobstructive temperature feedback is one developed for a prosthetic hand but not tested on human skin, where a multilayer heater and sensor array is laminated onto a prosthetic hand [41]. Without the use of effective temperature feedback, past stretchable heating devices have relied on the relationship between voltage and heat generated in order to maintain the heater at a desired temperature. However, heat transfer conditions vary from person to person and it is inaccurate to assume a consistent relationship between voltage and temperature if you wish to apply the same heater to multiple subjects. For example, changes in blood flow can cause changes in epidermal skin temperature [7].

Our group has developed a "cut-and-paste" method [42], in which stretchable patterns are cut out of ultrathin metal-polymer laminates and pasted to an adhesive substrate, allowing for cheaper, quicker and greener fabrication of tattoo-like sensors. This method also allows for easy integration of independent heaters and resistance temperature detectors (RTDs) on the same substrate. Using this method, we herein present an inexpensive, easy to fabricate and power-efficient programmable tattoo-like heating device which comprises a stretchable resistive heating element (RHE) of serpentine-shaped aluminum (Al) ribbons and a stretchable RTD of serpentine-shaped Au ribbons. The RTD is thin enough to not disturb the uniformity of the heat from the heating element. Included with this device is a customized proportional-integral-derivative (PID) control software which uses real time temperature feedback to control the heater and can maintain it at a target temperature over a large area of skin for extended periods of time.

2. Materials and Methods

The "cut-and-paste" manufacturing process of the stretchable tattoo-like heater is illustrated in Figure 1a, and a picture of the as-fabricated sample on 3M Tegaderm tape is offered in Figure 1b. As depicted in the first row of Figure 1a, the process began with placing a blanket 7 μm/13 μm Al/PET bilayer laminate (Neptco Inc., Pawtucket, RI, USA) smoothly on a thermal release tape (Semiconductor Equipment Corp., Moorpark, CA, USA) with Al facing up. A Silhouette mechanical cutter plotter was programmed to cut the designed seams on the bilayer within 3 min. Excessive Al/PET was removed once the thermal release tape (TRT) was heated and the remaining Al/PET ribbon was printed on a 3M Tegaderm tape with the Al side facing the Tegaderm and the bluish PET side facing outward. The 13-μm thick PET layer allows for increased mechanical integrity [43] and electrical insulation. The same process was repeated to cut and paste the stretchable RTD ribbon, which was made out of 100 nm/15 nm/13 μm Au/Cr/PET laminate, with the PET facing the Tegaderm and the Au facing outward, as illustrated by the second row of Figure 1a. This arrangement resulted in two layers of insulation between the Al RHE and the Au RTD at locations where the two intersected. Both the RHE and the RTD were cut into serpentine ribbons, which contributes to the stretchability and softness of the device. Specifically, the stretchability and softness of these serpentine ribbons can be maximized by fabricating their width to be as narrow as possible [37]. Due to the resolution of the Silhouette cutter, all ribbon widths were fixed to be 400 μm [42]. Although the resolution is far from photolithographic patterning technologies, the cost of time, materials and facilities is significantly reduced using the freeform cut-and-paste process because it does not require any chemicals, photomasks or cleanroom facilities. Moreover, while photolithographic process is limited to wafer scale, the patterning area of the cutter plotter can be as large as 30 cm. wide and a meter long.

Costs of previous epidermal electronic systems are dominated by fabrication processes such as spin coating, photolithography, wet and dry etching and transfer-printing. These methods require expensive clean-room fees and chemical purchases and they are also very time intensive. Using the "cut-and-paste" method allows for the fabrication of the presented device without any of those costs, making it significantly more cost effective than other similar epidermal electronics [38,42,44,45].

Finally, an ultrathin, ultrasoft double-sided tattoo adhesive was laid on top of the RHE and the RTD, providing a final layer of electrical insulation as well as increased adhesion between the skin and the patch. Snap button connectors were used to connect lead wires to both the RHE and the RTD (Supplementary Information, Figure S1).

(a)

(b)

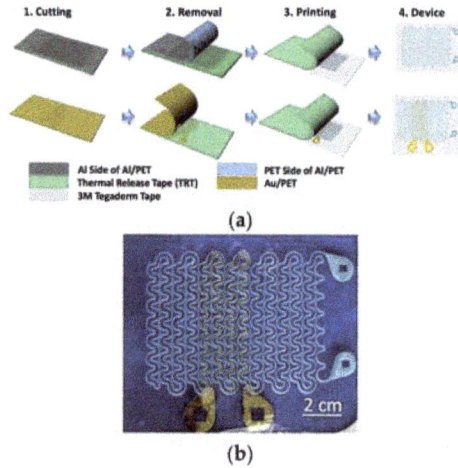

Figure 1. (a) Fabrication process used for heater and resistance temperature detector (RTD), shown for heater. Material is put on the thermal release tape (TRT) and cut with Silhouette cutter. TRT is heated, excess material is removed and remaining material is transferred to Tegaderm; (b) Complete device on tegaderm. Aluminum with blue polyimide backing forms the resistive heating element while Au/Cr 100/10 nm forms the resistance temperature detector.

The palm of a human subject's hand was chosen for the location to test the device on. The palms of the hand are glabrous skin surfaces, and heating them along with the soles of the feet can efficiently warm the body during anesthesia [7,8]. Presenting that the heater works effectively on the palms of the hand therefore demonstrates that perioperative warming is a feasible application of this device. When attached to the skin, this device conformed to the skin and deformed alongside it without mechanical resistance, as evidenced by Figure 2a,b. When a DC voltage of 5.1 V was applied across the Al RHE, it supplied an even amount of heat over the palm around the target temperature of 40 °C. There was minimum change in temperature during severe skin deformation such as hand clenching, as demonstrated by Figure 2c,d, which were taken by an infrared (IR) FLIR T620 camera (FLIR, Wilsonville, OR, USA). The University of Texas at Austin IRB protocol number for the human subject experiments was 2010-03-0050.

To calibrate the RTD, it was placed on an insulated hot plate and its resistance was compared against the temperature readings of two custom made type T thermocouples. The RTD exhibited the expected linear relationship [15] between resistance and temperature with a temperature coefficient of resistance (TCR) of 0.0025 °C^{-1} (Supplementary Information, Figure S2). To obtain the TCR under service condition, calibration of the RTD was also conducted on skin together with the RHE, and an IR camera was used for temperature measurement. In this set-up, the heat comes from the RHE instead of the hotplate and the RTD is in intimate contact with the actual heat sink—the skin. Due to these differences, we hypothesized that the TCR would be different from that measured on the hotplate. A schematic of the calibration set-up is depicted in (Figure 3a). The RHE was linked to a DC voltage supply (Mastech Linear Power Supply HY1803D, Pittsburgh, PA, USA) while the RTD was connected to a digital multimeter (DMM, NI Elvis II). Resistance readings were logged using the DMM and LabVIEW 2014. The device was covered with a fine layer of Johnson's Baby Powder to control its thermal radiation emissive properties [46]. The DC voltage supply was set to different voltages and the temperature and resistance of the RTD were measured simultaneously using the IR camera and the DMM, respectively. Temperature readings from the IR camera were logged using FLIR Tools+.

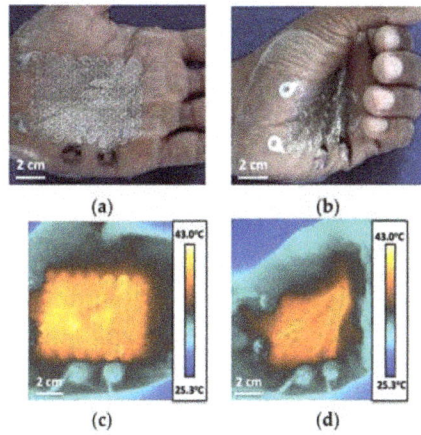

Figure 2. (**a**,**b**) Device conforms to hand and maintains its conformability during opening and closing; (**c**,**d**) Infrared (IR) images of the device powered with proportional-integral-derivative (PID) control as the hand is opened and closed. The PID controller automatically adjusts power output so the hand does not overheat when it closes.

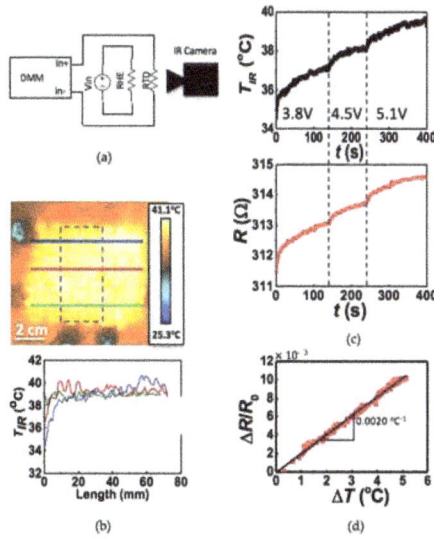

Figure 3. (**a**) Circuit diagram of set-up for calibration of RTD in situ. Heater is brought to different temperatures by adjusting Vin. Resistance and temperature are measured simultaneously using a digital multimeter (DMM) and IR camera, respectively; (**b**) Lateral heat distribution of heater. Blue, red and green lines on IR image mark the horizontal line across which temperature was measured for their respective red, blue and green plots. Temperature distribution is fairly uniform. Dotted purple line on IR image shows area that the IR camera calculated the average temperature for; (**c**) Average temperature of area marked by the dotted purple line in Figure 3b (top) and resistance of Au/Cr RTD as measured by DMM (bottom) each plotted across time as Vin was changed to 3.8 V, 4.5 V and finally 5.1 V; (**d**) The calibration curve for the RTD: $\Delta R/R_0$ of the RTD versus ΔT of the average temperature of the area around the RTD as marked by the dotted purple line in Figure 3b. The calibration constant, β, is marked and is equal to 0.000203

For safety purposes, the RHE-RTD calibration was first carried out on a glass slide, which has thermal properties similar to those of human skin [17]. The TCR was measured to be 0.0022 °C (Supplementary Information, Figure S3), which is slightly lower than that measured by the hotplate calibration. After ensuring the RHE behavior, a similar RHE-RTD calibration was performed on human skin. Figure 3b upper frame shows the IR image of the heater on human palm. It is evident that the temperature across the heater is fairly uniform over an area of 60 mm × 45 mm. The dotted black box indicates where the RTD resides. The IR temperature for the RTD calibration used in Figure 3c,d was obtained by averaging the temperature within this boxed area. It is clear in Figure 3b that the existence of the RTD does not affect the RHE or the temperature distribution. The three solid horizontal lines drawn across the heater mark the locations where the temperature is plotted as a function of distance along the lines in Figure 3b lower frame. Within the area covered by the RHE, temperature variation is between 38 °C and 40 °C. To continuously increase the temperature, the DC voltage was set to 3 increasing values: 3.8 V, 4.5 V and 5.1 V. Synchronously measured temperature and resistance versus time curves are provided in Figure 3c, which shows excellent alignment. Also visible in Figure 3c is the steady state average temperature of the RTD-area of the heater reaches when power is applied directly to the RHE. The average temperature to voltage ratio for the device was found to be 8.7 °C/V (Supplementary Information, Table S1). Plotting relative resistance change versus temperature change in Figure 3d, a linear fit with a TCR of 0.0020 °C^{-1} can be obtained. As expected, this is lower than the TCR found with the hotplate calibration (0.0025 °C^{-1}) or the glass substrate calibration (0.0022 °C^{-1}) due to the fact that the RTD is well conformed to human skin, beneath which blood flow can help mitigate the heat.

To verify the experimental findings, we ran a COMSOL simulation of the device heating human skin. The skin was modeled as a multilayer substrate made up of epidermis (0.1 mm thick), papillary dermis (0.7 mm thick), reticular dermis (0.8 mm thick), fat (2 mm thick) and muscle (16.4 mm thick), each with different thermophysical properties taken from literature [47]. No blood perfusion effects were included in the model. Ambient radiation from the RHE and convective cooling between the RHE and the environment were taken into consideration. With the environment temperature set at 15 °C and the core temperature set at 37 °C, the skin surface temperature stabilized at 34.4 °C when the heater was off. The effective electrical conductivity of the RHE was calibrated by setting the maximum temperature to be 41.4 °C when the applied voltage was 5.1 V. Using a Joule heating model for the RHE and a heat transfer model for the other components of the device and the skin, the modeled temperature distribution across the skin was found under transient and equilibrium states. Figure 4a,b displays the top and 3D cross-sectional views of the temperature distribution within the skin under equilibrium while the heater was on. Figure 4c plots temperature distributions along the three lines drawn on the left frame of Figure 4a where the blue, red and green curves correspond to the blue, red and green lines, respectively. The close agreement between Figures 3b and 4c validates the COMSOL model and gives more credit to the simulated equilibrium temperature distribution in the skin along the depth direction (as indicated by the black arrow in Figure 4a right frame), as plotted in Figure 4c. When the heater is off (dashed curve), skin surface temperature is 34.4 °C. As the depth increases, the curve approaches the core temperature of 37 °C. When the heater is on (solid curve), skin surface is heated to 41.4 °C. The temperature gradually decays to 37 °C as we go deep into the skin. The slight kinks in the curves are due to the change of the thermophysical properties of the different layers of human skin.

(a)

(b)

(c)

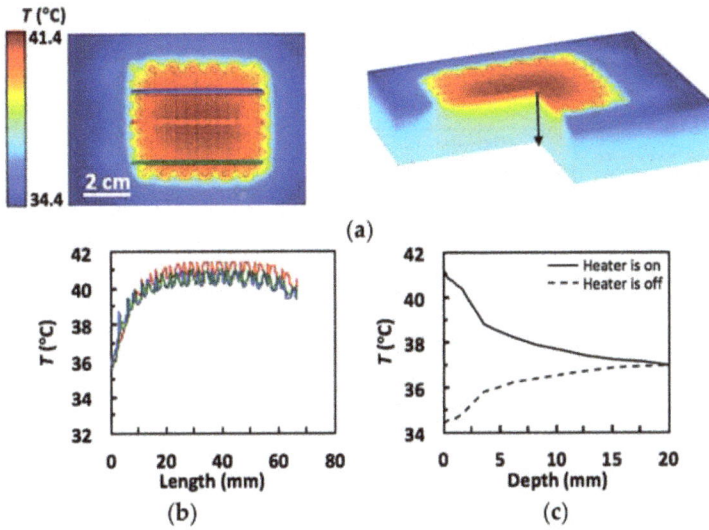

Figure 4. (a) COMSOL thermal simulation results (left: top view; right: 3D view); (b) Lateral heat distribution of heater. Blue, red and green lines on simulation image (left) mark the horizontal line across which temperature was collected for their respective red, blue and green plots; (c) Vertical heat distribution of skin from the black line on simulation image (right).

After calibrating the RTD and characterizing the RHE, we were able to establish a real time PID feedback control as illustrated by the diagram in Figure 5a. The purpose of this system was to demonstrate the functionality of the tattoo-like heater itself and was thus built with wires connecting the heater to a data acquisition (DAQ) unit and a PC. The system can be made to be wireless in the future by integrating it with a microcontroller unit (MCU), a Bluetooth low energy (BLE) chip and a rechargeable battery on a miniature printed circuit board (PCB). The DC power to the RHE was routed through an Omron DC-DC relay (G3CN) which was controlled by a computer using an output DAQ (NI USB-6009). The computer ran a LabVIEW program which controlled the temperature of the RHE using pulse width modulation (PWM). The RTD was connected to the DMM of an NI Elvis II, which measured the RTD's resistance and sent the readings to the LabVIEW program in real time. The LabVIEW program converted the resistance readings into temperature using the following equation:

$$T = T_0 + \frac{\Delta R}{0.002 R_0},$$

(1)

where the initial resistance R_0 was measured at the room temperature T_0 and the coefficient $0.0020\,^{\circ}\text{C}^{-1}$ was the TCR obtained from the calibration on human skin in Figure 3d. The PID program then used the real-time temperature feedback, along with a desired temperature set point, to determine how to control the relay and thereby the PWM of the RHE. This allowed the program to keep the heater at a set temperature or to adjust to a new temperature when demanded.

(a)

(b)

(c)

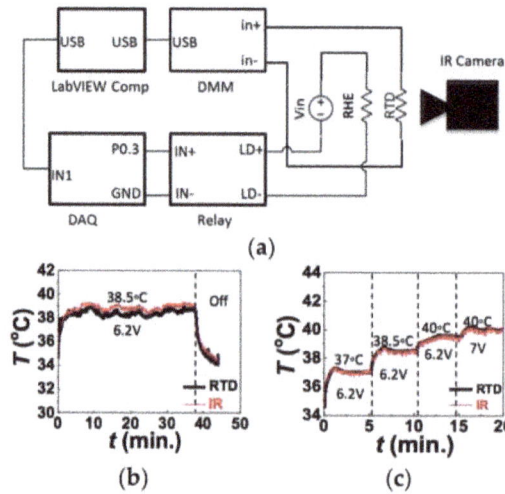

Figure 5. (a) Circuit diagram of set up for operating heater with PID control. DMM measures resistance of RTD and feeds it into a computer with LabVIEW, The LabVIEW program calculates the temperature of the RTD using the RTD's starting temperature, starting resistance and calibration constant. It then uses a PID algorithm to calculate the optimal duty cycle for PWM of the heater given the heater's current temperature and the set point temperature for the heater. The LabVIEW program then uses the data acquisition unit (DAQ) to switch the relay on and off with the determined duty cycle, thus controlling how much total power is fed to the heater; (b) Temperature of the heater versus time measured with both the RTD and the IR camera as the heater is turned on at a set point of 38.5 °C and then turned off. Heater is able to maintain set point temperature for an extended period of time; (c) Temperature of the heater versus time measured with both the RTD and the IR camera as the set point of the heater is changed while the voltage remains constant. At 40 °C, 6.2 V is not sufficient for the heater to reach the set point, so the voltage is increased to 7 V, at which point the heater is able to reach and maintain a temperature of 40 °C.

3. Results and Discussion

First, to test if the device could effectively maintain a target temperature the DC voltage supply was set to 6.2 V and the temperature was set to 38.5 °C. The device was able to maintain a constant temperature of 38.5 °C for over 30 min on human skin until the heater was completely turned off to finish the experiment as shown in Figure 5b. The heater reached the target temperature within 3 min and the error between the target temperature and the device's temperature never reached more than a degree (Supplementary Information, Figure S4). The temperature readings of the RTD (black curve) was also verified by the IR camera results (red curve).

To test if the device could self-adjust when set temperature changes, we conducted an experiment with multiple set temperatures (37 °C, 38.5 °C, 40 °C) while the voltage supply was kept constant at 6.2 V (Figure 5c). For the first two temperatures (Stages I and II), the device was able to reach the set temperatures and to maintain them at a steady state. In switching between these temperatures, no changes were made except changing set point in the LabVIEW program. When the voltage was kept at 6.2 V and the target temperature was set to 40 °C, which is marked as Stage III, the actual skin surface temperature was not able to reach 40 °C. It could only reach up to 39 °C. This indicates insufficient power supply even when the duty cycle of the PWM reached 100%. Therefore, the maximum steady state temperature a heater can reach at 6.2 V with this set-up and under these circumstances is 39 °C, demonstrating a temperature to voltage ratio of 6.19 V/°C. This is lower than what was observed when the power was applied directly to the RHE and not routed through the relay. This indicates that some power may be lost as the electricity passes through the relay and the wires thereto. We therefore increased the voltage to 7 V and the skin surface was then successfully heated to 40 °C, as in Stage IV. Again, the temperature measured by the RTD (black) and the IR camera (red) are well matched. This experiment demonstrates that when given a sufficient voltage supply, the stretchable tattoo-like heater can automatically reach, maintain and change between desired temperatures without any manual adjustment of the voltage.

Due to the negligible stiffness of serpentine ribbons [48,49], the mechanical stiffness of our tattoo-like heater is dominated by the supporting Tegaderm tape, whose Young's modulus was measured to be 7 MPa [42]. The effects of strain and skin deformation on our stretchable heater are discussed in Supplementary Information, Figures S5 and S6. Figure S5 indicates that our RTD can survive more than 70% tensile strain but its resistance is unfortunately slightly sensitive to strain. Due to such strain effects, Figure S6 indicates that the RTD temperature is not accurate when the hand closes but its measurement can recover when the hand restores its original configuration. With the PID control, the RHE may under heat the skin due to the falsely perceived increase of RTD temperature during hand closure, which will not cause any skin burn.

To evaluate the power consumption of the tattoo-like heater, the duty cycles at different set temperatures were investigated. The device was placed on a human palm with PID control. The top frame in Figure 6a plots the actual skin temperature measured by the RTD versus time and the labels are again voltage supply and set temperature. Some small overshooting of the temperature occasionally happens at the points where voltage is changed but they are small and quickly rectified by the controller. The device is not expected to undergo step changes in voltages during real life application but these experiments show that the controller can react appropriately to those step changes should they occur. At each set temperature, the steady state duty cycle was recorded. The middle frame of Figure 6a shows the duty cycle versus time plot. The numbers mark the plateaus where the device was considered to have reached steady state. The duty cycle for each set temperature was calculated as the average of the duty cycle readings at these plateaus. The following equation was then used for power calculation:

$$P = \frac{D}{100\%} \times \frac{V^2}{R},\tag{2}$$

where D is the duty cycle, V is the voltage supplied to the RHE and R is the resistance of the RHE.

If we define power density to be the power delivered to the skin per unit area of the RHE, power density can be calculated through:

$$\text{Power Density} = P/A,\tag{3}$$

where A represents the total area of the heater, which is 38.7 cm^2 for our RHE. Plotting power density versus the corresponding temperature as red markers in the bottom frame of Figure 6a, a linear relation can be fitted. The slope of this linear curve is defined as the specific power flow (SPF), which represents power density normalized to the applied thermal driving potential, that is, temperature difference. The SPF of our stretchable tattoo-like heater is estimated to be 0.846 mW/(cm$^2 \cdot$°C), which means that to heat up a 1 cm^2 area of this specific human palm by 1 °C would consume a power of 0.846 mW.

Considering convection and radiation between the heater and the ambient environment, it is inaccurate to assume that all the heat generated by the RHE completely goes into the skin. To obtain a more accurate estimation of the specific power flow into the skin, the entire experiment of Figure 6a was repeated in Figure 6b but with insulation over the heater. A 4 cm thick layer of foam, which is a well-known heat insulator, was taken from a delivery package and applied over the heater on the palm to minimize heat loss into the environment. With this heat insulating foam, the SPF was found to be 0.784 mW/(cm$^2 \cdot$°C) as given in Figure 6b bottom frame, which is 7.33% lower compared with that of the exposed heater (0.846 mW/(cm$^2 \cdot$°C)). This result indicates that about 7.33% of the heat generated by the RHE was lost to the environment when the RHE was exposed to air.

To compare our stretchable tattoo-like heater with other stretchable heaters in the literature, we summarized their materials, substrates, power densities and SPFs in Table 1. In cases where power supplied without using PWM, the value of *D* in Equation (2) was set to 100%. It is evident that the SPF of our tattoo-like heater is the lowest among the stretchable heaters directly applied on human skin, which is an active heat sink in comparison with air, polydimethylsiloxane (PDMS) and glass. Moreover, our device is one of the first to implement real time feedback control for stretchable heaters on human skin.

If adopted for commercialization, our device could very feasibly be converted into a wireless portable device with a battery-operated microcontroller as has been demonstrated for other joule heating devices reported in the Table 1 [5,30,31].

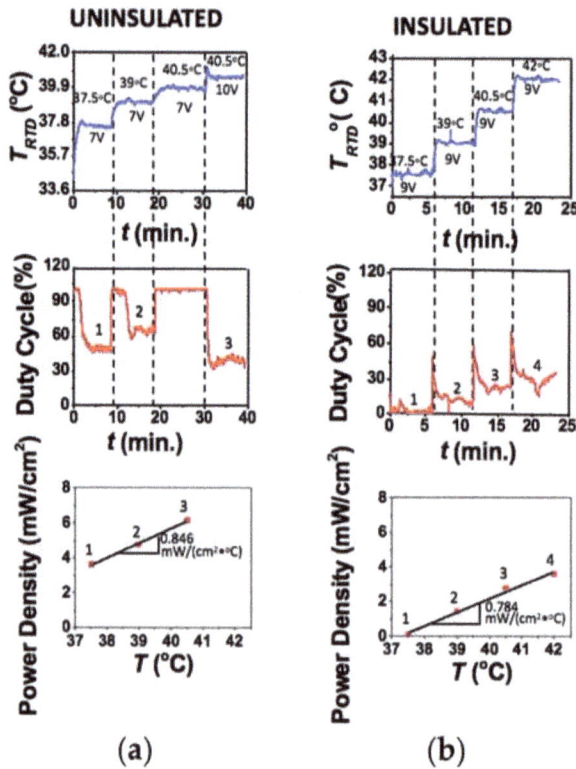

Figure 6. (a) Plot of heater temperature versus time as the set points and voltages are changed, followed by a plot of the corresponding duty cycle versus time. The average of the steady state duty cycles marked 1, 2 and 3 were used to calculate the power densities plotted below marked 1, 2 and 3, respectively, at different temperatures; (b) The same plots as figure A except the heater is insulated with a piece of foam.

Table 1. Compiled information about different stretchable heaters.

Ref.	T Sensor on Site	Feedback Control	Target T (°C)	Resistive Heating Element	Substrate	V (V)	R (Ω)	Area (cm²)	P (W)	Power Density (mW/cm²)	SPF mW/(cm²·°C)
Our device	Yes	Yes	43	Al (9 μm)	Skin	10	17.2	38.7	2.38	61.44	0.78
[5]	No	No	43	Ag NW/SBS (18/82)	Skin	3.7	~2	91	~7	~80	0.9
[9]	Yes	Yes	40	Au (190 nm)	Pig Skin	12	95.9	<2.3	1.5	>652	>40 *
[15]	Yes	No	ΔT = 6	Au	Skin	–	–	0.64	0.01	20.31	3
[30]	Yes	Yes	250	Ag/Ethylene Glycol (50 wt %)	50 μm PI	4.5	0.75	–	27	650	3
[31]	No	No	40	CuZr nanotrough network	Skin	1.7	3.9	1.24	0.74	600	35
[39]	No	No	39	Ag NW /PDMS 132 mg m⁻²	Air	4	50	38.5	0.32	8.31	0.6
[39]	No	No	56	Ag NW /PDMS 396 mg m⁻²	Air	5	15	38.5	1.67	43.29	1
[41]	Yes	No	37	Cr/Au 7/70 nm	PDMS	4.4	550	9	0.04	3.91	0.8
[40]	Yes	Yes	31	PE Yarn with CuNW coating	Skin	1.4	1.66	30.16	1.18	39.12	5

* Assuming power for glass experiments was same as power for pig skin experiments.

4. Conclusions

A low cost, low power consumption stretchable tattoo-like heater was fabricated to reliably warm the skin surface to a target temperature. The device combines a stretchable RHE and a stretchable RTD into a single unit through the cost and time effective "cut-and-paste" fabrication method. The device is thin (60 μm thick) and soft (7.4 MPa modulus) so that it can conform to the complex 3-D surface of palms and can deform and remain attached during hand flexure without perceivable mechanical resistance. The RHE is able to reach set temperatures with relatively even distribution, including during hand movement. The RTD can monitor the real-time temperature of the palm accurately, as verified by simultaneous IR measurements. Through PID temperature feedback control, the device is able to maintain set temperatures for extended periods of time and can automatically adjust to a different temperature if the set point is changed on the controller. The SPF of our device is comparable with reported stretchable skin-mounted heaters. Its simple circuit and program can easily be downscaled to a battery powered printed circuit board (PCB) and microcontroller, giving it potential for point-of-care applications.

Supplementary Materials: The following are available online at http://www.mdpi.com/2072-666X/9/4/170/s1, Figure S1: Snap button connections, Figure S2: RTD calibrated on hotplate, Figure S3: RTD calibrated on glass with RHE, Figure S4: Error from set point during extended palm heating with device with PID control. Figure S5: Stretchability test, Figure S6: Effect of skin deformation, Table S1: Steady state temperatures for different voltages and temperature to voltage ratios, Video S1: Hand Clench Test.

Acknowledgments: This work is supported by the Office of Naval Research Young Investigator Award under Contract N00014-16-1-2044 to N.L. and National Science Foundation Grant BET1250659 to K.R.D.

Author Contributions: A.S., E.H., S.Y. and N.L. conducted device design, fabrication, calibration and testing. A.M. constructed the initial PID control algorithm and A.S. later revised it for this project. S.Q. performed the COMSOL finite element modeling. N.L. and K.R.D. supervised and coordinated the project. A.S., N.L., S.Q. and K.R.D. wrote the paper.

Conflicts of Interest: The authors declare no conflict of interest.

References

1. Petrofsky, J.; Berk, L.; Bains, G.; Khowailed, I.A.; Hui, T.; Granado, M.; Laymon, M.; Lee, H. Moist heat or dry heat for delayed onset muscle soreness. *J. Clin. Med. Res.* **2013**, *5*, 416–425. [CrossRef] [PubMed]
2. Nadler, S.F.; Weingand, K.; Kruse, R.J. The physiologic basis and clinical applications of cryotherapy and thermotherapy for the pain practitioner. *Pain Physician* **2004**, *7*, 395–399. [PubMed]
3. Brosseau, L.; Yonge, K.; Welch, V.; Marchand, S.; Judd, M.; Wells, G.; Tugwell, P. Thermotherapy for treatment of osteoarthritis. *Cochrane Database Syst. Rev.* **2003**, *4*. [CrossRef]
4. Lehmann, J. *Therapeutic Heat and Cold*, 4th ed.; Williams & Wilkins: Baltimore, MD, USA, 1990.
5. Choi, S.; Park, J.; Hyun, W.; Kim, J.; Kim, J.; Lee, Y.B.; Song, C.; Hwang, H.J.; Kim, J.H.; Hyeon, T.; et al. Stretchable heater using ligand-exchanged silver nanowire nanocomposite for wearable articular thermotherapy. *ACS Nano* **2015**, *9*, 6626–6633. [CrossRef] [PubMed]
6. Sessler, D.I.; Rubinstein, E.H.; Moayeri, A. Physiologic responses to mild perianesthetic hypothermia in humans. *Anesthesiology* **1991**, *75*, 594–610. [CrossRef] [PubMed]
7. Diller, K.R. Heat transfer in health and healing. *J. Heat Transf.* **2015**, *137*, 1030011–10300112. [CrossRef] [PubMed]
8. Grahn, D.; Brock-Utne, J.G.; Watenpaugh, D.E.; Heller, H.C. Recovery from mild hypothermia can be accelerated by mechanically distending blood vessels in the hand. *J. Appl. Physiol.* **1998**, *85*, 1643–1648. [CrossRef] [PubMed]
9. Son, D.; Lee, J.; Qiao, S.; Ghaffari, R.; Kim, J.; Lee, J.E.; Song, C.; Kim, S.J.; Lee, D.J.; Jun, S.W.; et al. Multifunctional wearable devices for diagnosis and therapy of movement disorders. *Nat. Nanotechnol.* **2014**, *9*, 397–404. [CrossRef] [PubMed]
10. Bagherifard, S.; Tamayol, A.; Mostafalu, P.; Akbari, M.; Comotto, M.; Annabi, N.; Ghaderi, M.; Sonkusale, S.; Dokmeci, M.R.; Khademhosseini, A. Dermal patch with integrated flexible heater for on demand drug delivery. *Adv. Healthc. Mater.* **2016**, *5*, 175–184. [CrossRef] [PubMed]

11. Petrofsky, J.S.; Laymon, M.; Lee, H. Effect of heat and cold on tendon flexibility and force to flex the human knee. *Med. Sci. Monit. Int. Med. J. Exp. Clin. Res.* **2013**, *19*, 661–667. [CrossRef]

12. Michlovitz, S.; Hun, L.; Erasala, G.N.; Hengehold, D.A.; Weingand, K.W. Continuous low-level heat wrap therapy is effective for treating wrist pain. *Arch. Phys. Med. Rehabil.* **2004**, *85*, 1409–1416. [CrossRef] [PubMed]

13. Wang, S.; Li, M.; Wu, J.; Kim, D.-H.; Lu, N.; Su, Y.; Kang, Z.; Huang, Y.; Rogers, J.A. Mechanics of epidermal electronics. *J. Appl. Mech.* **2012**, *79*, 031022. [CrossRef]

14. Wang, L.; Lu, N. Conformability of a thin elastic membrane laminated on a soft substrate with slightly wavy surface. *J. Appl. Mech.* **2016**, *83*, 041007. [CrossRef]

15. Webb, R.C.; Bonifas, A.P.; Behnaz, A.; Zhang, Y.; Yu, K.J.; Cheng, H.; Shi, M.; Bian, Z.; Liu, Z.; Kim, Y.-S.; et al. Ultrathin conformal devices for precise and continuous thermal characterization of human skin. *Nat. Mater.* **2013**, *12*, 938–944. [CrossRef] [PubMed]

16. Cartmell, J.V.; DeRosa, J.F. Capacitively Coupled Indifferent Electrode. U.S. Patent 4,669,468, 2 June 1987.

17. Roselli, R.J.; Diller, K.R. *Biotransport: Principles and Applications*, 2011 ed.; Springer: New York, NY, USA, 2011; ISBN 978-1-4419-8118-9.

18. Tuckey, J. Forced-air warming blanket and surgical access. *Anaesthesia* **1999**, *54*, 97–98. [CrossRef] [PubMed]

19. Kimberger, O.; Held, C.; Stadelmann, K.; Mayer, N.; Hunkeler, C.; Sessler, D.I.; Kurz, A. Resistive polymer versus forced-air warming: Comparable heat transfer and core rewarming rates in volunteers. *Anesth. Analg.* **2008**, *107*, 1621–1626. [CrossRef] [PubMed]

20. Rogers, J.A.; Someya, T.; Huang, Y. Materials and mechanics for stretchable electronics. *Science* **2010**, *327*, 1603–1607. [CrossRef] [PubMed]

21. Kim, D.-H.; Ghaffari, R.; Lu, N.; Rogers, J.A. Flexible and stretchable electronics for biointegrated devices. *Annu. Rev. Biomed. Eng.* **2012**, *14*, 113–128. [CrossRef] [PubMed]

22. Suo, Z. Mechanics of stretchable electronics and soft machines. *MRS Bull.* **2012**, *37*, 218–225. [CrossRef]

23. Nassar, J.M.; Rojas, J.P.; Hussain, A.M.; Hussain, M.M. From stretchable to reconfigurable inorganic electronics. *Extreme Mech. Lett.* **2016**, *9*, 245–268. [CrossRef]

24. Sekitani, T.; Noguchi, Y.; Hata, K.; Fukushima, T.; Aida, T.; Someya, T. A rubberlike stretchable active matrix using elastic conductors. *Science* **2008**, *321*, 1468–1472. [CrossRef] [PubMed]

25. Jung, S.; Kim, J.H.; Kim, J.; Choi, S.; Lee, J.; Park, I.; Hyeon, T.; Kim, D.-H. Reverse-micelle-induced porous pressure-sensitive rubber for wearable human–machine interfaces. *Adv. Mater.* **2014**, *26*, 4825–4830. [CrossRef] [PubMed]

26. Kim, Y.; Zhu, J.; Yeom, B.; Di Prima, M.; Su, X.; Kim, J.-G.; Yoo, S.J.; Uher, C.; Kotov, N.A. Stretchable nanoparticle conductors with self-organized conductive pathways. *Nature* **2013**, *500*, 59–63. [CrossRef] [PubMed]

27. Chun, K.-Y.; Oh, Y.; Rho, J.; Ahn, J.-H.; Kim, Y.-J.; Choi, H.R.; Baik, S. Highly conductive, printable and stretchable composite films of carbon nanotubes and silver. *Nat. Nanotechnol.* **2010**, *5*, 853–857. [CrossRef] [PubMed]

28. Ma, R.; Lee, J.; Choi, D.; Moon, H.; Baik, S. Knitted fabrics made from highly conductive stretchable fibers. *Nano Lett.* **2014**, *14*, 1944–1951. [CrossRef] [PubMed]

29. Park, M.; Im, J.; Shin, M.; Min, Y.; Park, J.; Cho, H.; Park, S.; Shim, M.-B.; Jeon, S.; Chung, D.-Y.; et al. Highly stretchable electric circuits from a composite material of silver nanoparticles and elastomeric fibres. *Nat. Nanotechnol.* **2012**, *7*, 803–809. [CrossRef] [PubMed]

30. Jang, J.; Hyun, B.G.; Ji, S.; Cho, E.; An, B.W.; Cheong, W.H.; Park, J.-U. Rapid production of large-area, transparent and stretchable electrodes using metal nanofibers as wirelessly operated wearable heaters. *NPG Asia Mater.* **2017**, *9*, e432. [CrossRef]

31. An, B.W.; Gwak, E.-J.; Kim, K.; Kim, Y.-C.; Jang, J.; Kim, J.-Y.; Park, J.-U. Stretchable, transparent electrodes as wearable heaters using nanotrough networks of metallic glasses with superior mechanical properties and thermal stability. *Nano Lett.* **2016**, *16*, 471–478. [CrossRef] [PubMed]

32. Moon, G.D.; Lim, G.-H.; Song, J.H.; Shin, M.; Yu, T.; Lim, B.; Jeong, U. Highly stretchable patterned gold electrodes made of Au nanosheets. *Adv. Mater.* **2013**, *25*, 2707–2712. [CrossRef] [PubMed]

33. Gray, D.S.; Tien, J.; Chen, C.S. High-conductivity elastomeric electronics. *Adv. Mater.* **2004**, *16*, 393–397. [CrossRef]

34. Li, T.; Suo, Z.; Lacour, S.P.; Wagner, S. Compliant thin film patterns of stiff materials as platforms for stretchable electronics. *J. Mater. Res.* **2005**, *20*, 3274–3277. [CrossRef]
35. Xu, S.; Zhang, Y.; Cho, J.; Lee, J.; Huang, X.; Jia, L.; Fan, J.A.; Su, Y.; Su, J.; Zhang, H.; et al. Stretchable batteries with self-similar serpentine interconnects and integrated wireless recharging systems. *Nat. Commun.* **2013**, *4*, 1543. [CrossRef] [PubMed]
36. Fan, J.A.; Yeo, W.-H.; Su, Y.; Hattori, Y.; Lee, W.; Jung, S.-Y.; Zhang, Y.; Liu, Z.; Cheng, H.; Falgout, L.; et al. Fractal design concepts for stretchable electronics. *Nat. Commun.* **2014**, *5*, 3266. [CrossRef] [PubMed]
37. Yang, S.; Ng, E.; Lu, N. Indium Tin Oxide (ITO) serpentine ribbons on soft substrates stretched beyond 100%. *Extreme Mech. Lett.* **2015**, *2*, 37–45. [CrossRef]
38. Kim, D.-H.; Lu, N.; Ma, R.; Kim, Y.-S.; Kim, R.-H.; Wang, S.; Wu, J.; Won, S.M.; Tao, H.; Islam, A.; et al. Epidermal electronics. *Science* **2011**, *333*, 838–843. [CrossRef] [PubMed]
39. Hong, S.; Lee, H.; Lee, J.; Kwon, J.; Han, S.; Suh, Y.D.; Cho, H.; Shin, J.; Yeo, J.; Ko, S.H. Highly stretchable and transparent metal nanowire heater for wearable electronics applications. *Adv. Mater.* **2015**, *27*, 4744–4751. [CrossRef] [PubMed]
40. Cheng, Y.; Zhang, H.; Wang, R.; Wang, X.; Zhai, H.; Wang, T.; Jin, Q.; Sun, J. Highly stretchable and conductive copper nanowire based fibers with hierarchical structure for wearable heaters. *ACS Appl. Mater. Interfaces* **2016**, *8*, 32925–32933. [CrossRef] [PubMed]
41. Kim, J.; Lee, M.; Shim, H.J.; Ghaffari, R.; Cho, H.R.; Son, D.; Jung, Y.H.; Soh, M.; Choi, C.; Jung, S.; et al. Stretchable silicon nanoribbon electronics for skin prosthesis. *Nat. Commun.* **2014**, *5*, 5747. [CrossRef] [PubMed]
42. Yang, S.; Chen, Y.-C.; Nicolini, L.; Pasupathy, P.; Sacks, J.; Su, B.; Yang, R.; Sanchez, D.; Chang, Y.-F.; Wang, P.; et al. "Cut-and-Paste" Manufacture of multiparametric epidermal sensor systems. *Adv. Mater.* **2015**, *27*, 6423–6430. [CrossRef] [PubMed]
43. Lu, N.; Wang, X.; Suo, Z.; Vlassak, J. Metal films on polymer substrates stretched beyond 50%. *Appl. Phys. Lett.* **2007**, *91*, 221909. [CrossRef]
44. Lu, N.; Kim, D.-H. Flexible and stretchable electronics paving the way for soft robotics. *Soft Robot.* **2013**, *1*, 53–62. [CrossRef]
45. Kim, J.; Banks, A.; Cheng, H.; Xie, Z.; Xu, S.; Jang, K.I.; Lee, J.W.; Liu, Z.; Gutruf, P.; Huang, X.; et al. Epidermal electronics with advanced capabilities in near-field communication. *Small* **2015**, *11*, 906–912. [CrossRef] [PubMed]
46. Methods of Increasing Emissivity in the Infrared Spectrum. Available online: http://www.optotherm.com/emiss-increasing.htm (accessed on 2 May 2016).
47. Cetingül, M.P.; Herman, C. A heat transfer model of skin tissue for the detection of lesions: Sensitivity analysis. *Phys. Med. Biol.* **2010**, *55*, 5933–5951. [CrossRef] [PubMed]
48. Widlund, T.; Yang, S.; Hsu, Y.-Y.; Lu, N. Stretchability and compliance of freestanding serpentine-shaped ribbons. *Int. J. Solids Struct.* **2014**, *51*, 4026–4037. [CrossRef]
49. Yang, S.; Qiao, S.; Lu, N. Elasticity solutions to nonbuckling serpentine ribbons. *J. Appl. Mech.* **2017**, *84*, 021004. [CrossRef]

micromachines

MDPI

Review

Tunable Adhesion for Bio-Integrated Devices

Zhaozheng Yu [1] and Huanyu Cheng [1,2,*]

[1] Department of Engineering Science and Mechanics, The Pennsylvania State University, University Park,
     State College, PA 16802, USA; zqy5106@psu.edu
[2] Materials Research Institute, The Pennsylvania State University, University Park, State College,
     PA 16802, USA
*   Correspondence: huanyu.cheng@psu.edu; Tel.: +1-(814)-863-5945

Received: 28 September 2018; Accepted: 16 October 2018; Published: 18 October 2018

Abstract: With the rapid development of bio-integrated devices and tissue adhesives, tunable adhesion to soft biological tissues started gaining momentum. Strong adhesion is desirable when used to efficiently transfer vital signals or as wound dressing and tissue repair, whereas weak adhesion is needed for easy removal, and it is also the essential step for enabling repeatable use. Both the physical and chemical properties (e.g., moisture level, surface roughness, compliance, and surface chemistry) vary drastically from the skin to internal organ surfaces. Therefore, it is important to strategically design the adhesive for specific applications. Inspired largely by the remarkable adhesion properties found in several animal species, effective strategies such as structural design and novel material synthesis were explored to yield adhesives to match or even outperform their natural counterparts. In this mini-review, we provide a brief overview of the recent development of tunable adhesives, with a focus on their applications toward bio-integrated devices and tissue adhesives.

Keywords: bio-integrated devices; tissue adhesives; tunable adhesion; dry/wet conditions; soft biological tissue

---

1. Introduction

Although adhesion has long been studied, early efforts focused on the contact between stiff materials [1]. Due to the emerging interest in reconfigurable systems [2] and bio-integrated devices [3,4], adhesion that involves a soft material with different levels of adhesion strength or even a tunable range started attracting attention. Soft materials of interest range from synthetic polymers to biological tissues [5]. Adhesion involving soft materials could be affected by the surface structure/morphology, the deformation of soft materials, and wet/dry conditions, among many others [6–8]. The strategies to design and achieve various levels of adhesion strength can be achieved through structural designs or material innovations. The rapid development in both these classes is greatly promoted by bio-inspiration from several marvelous animals (e.g., gecko, octopus, and mussel), which shed light on the effects of surface roughness, the directionality of the adhesive, and surface chemistry [6,9–11]. Although adhesion is greatly modulated by the properties of the adhesive layer, it is also affected by the target substrates due to interaction at the adhesive–substrate interface; thus, the adhesive has to be specifically designed for each application [12].

When it comes to adhesion to biological tissues, tunable adhesion is of great importance. For instance, strong adhesion to the wound edge is expected in a tissue adhesive to suture the wound [13]. Upon completion of wound healing, a weak adhesion is then desirable for easy removal of the adhesive. Tunable adhesion is also one essential step for realizing repeatable use, as easily removed adhesive could be sanitized and prepared for further use. Current commercial tissue adhesives such as Dermabond® [14] are designed for one-time use. However, when combined with tunable properties, tissue adhesives can be used as an alternative to surgical sutures in clinical practice to eliminate the need for stitch removal. As a decrease in adhesion is observed following multiple uses of the adhesive, strategies to minimize such a decrease need to be explored [15].

When integrated with functional sensing components and actuators, soft adhesives could provide a natural interface to enable bio-integrated devices (e.g., flexible displays [16–18], wellness monitors [19–23], and therapeutic devices [24,25]). The soft nature of the adhesive could minimize discomfort, while strong adhesion improves signal transfer through an enhanced signal-to-noise ratio [26]. Reversible adhesion also allows multiple uses of bio-integrated electronics. Moreover, the capability to modulate interfacial adhesion finds application in the development of transfer printing [27–30], which is widely used for the assembly of heterogeneous materials [31] with applications from wearable devices [3,32–34] to biodegradable sensors [35–39].

In this mini-review, we firstly provide a brief overview of the structural design for adhesives with applications mostly in the dry environment. As extensive review articles exist for dry adhesives [8,40–42], only selected key developments are highlighted here. Next, we discuss material innovations for using adhesives in the wet environment, which are largely based on bio-inspiration from mussels. As special considerations have to be given to the application of adhesives on biological tissue surfaces, we then highlight several recently developed techniques for such applications.

2. Structural Design for Dry Adhesion

Due to its remarkable ability to climb rapidly up a variety of vertical surfaces (Figure 1A (i)), the gecko inspired researchers to uncover the underlying mechanisms behind its significant enhanced, highly robust and repeatable, and reversible adhesion. Observation of the pad area (Figure 1A (ii)) shows nearly 500,000 keratin setae (pillars) (Figure 1A (iii)), with each seta consisting of branches of spatulas that are approximately 200 nm in diameter and 20–60 µm in length (Figure 1A (iv)) [11]. Experimental evidence confirmed that the dry adhesion of gecko setae results from van der Waals forces rather than mechanisms associated with a high surface polarity such as capillary adhesion [43], which indicates that the exceptional adhesion is merely a result of the size and shape of the setae tips. The direct observation of the van der Waals interaction indicates that the adhesion is not affected by the surface chemistry, and repeatable use is possible [40]. In order to reveal the role of the van der Waals interaction on the enhanced adhesion observed in the gecko pad [41,44,45], an array of biomimetic microscopic fibrils on an elastic support was created [41,46]. In direct contrast to a flat surface that only has limited contact to the target substrate with a microscale surface roughness, the array of fibrils with a high aspect ratio in a dense arrangement [47] was observed to form intimate contact with the target substrate due to its low effective Young's modulus and increased effective contact area, especially when a preload was applied (Figure 1B) [48,49]. The principle of contact mechanics was further applied to illustrate that contact splitting (i.e., reducing the radius of the fibril) yielded substantially improved adhesion, and the scaling was found to be applicable to animals differing in weight by six orders of magnitude (Figure 1C) [45,47,50]. The use of soft polymers in most biomimetic systems helps increase the adhesion, but their tacky nature also makes them more susceptible to particulate fouling; thus, a hydrophobic surface with the capacity for self-cleaning is desired. In fact, a fibrillar adhesive can partially transfer particles in a certain size range from its surface to the clean substrate and recover ca. one-third of its shear adhesion [51], as observed in gecko setae [52]. In the practical application where defects commonly exist, the fibrillar structure also localizes the contact failure at individual fibrils and minimizes the effect on contact adhesion, thereby increasing the defect tolerance. Moreover, the adhesion is also affected by the underlying supports. By peeling a polydimethylsiloxane (PDMS) substrate patterned with different hexagonal arrays of cylindrical pillars (to mimic fibrils) from an acrylic adhesive, the enhancement in the adhesion was shown to be more than the increase of the contact area, and this was attributed to the deformation of the underlying support [6]. The fibrillary structure can also be used directly beneath the existing viscoelastic adhesive film (e.g., pressure-sensitive adhesives) to change the dissipative crack trapping and the stress field in the viscoelastic layer for enhanced adhesion [53].

Due to the need for locomotion [54], a reversible adhesion is desirable, and animals such as geckos are observed to use direction to switch from strong to weak adhesion [55]. This direction-dependent adhesion is attributed to the angled fibrils on the gecko's foot, as evidenced by 20° cryo-SEM imaging [56]. While several methods were explored to fabricate vertical structures (e.g., e-beam lithography [57], nano-molding [58], constructing polymers from stiff thermoplastic [59], nano-drawing of stretched polymers [60], and growth of carbon nanotubes [61,62]), it is a challenge to obtain angled structures with high resolution and high aspect ratio, though several attempts were made (e.g., directional exposure in the lithography [63,64], deforming the shape memory polymer of vertical structures from soft lithography [65], post directional e-beam exposure of Pt-coated vertical polyurethane acrylate (PUA) nanohairs [66], and direct laser writing [67]). In another effort to address this challenge, an angled etching technique was developed, where a Faraday cage introduced in the conventional plasma etching system allows vertical movement of ions to induce an angled etching to the silicon substrate that is placed on an inclined stage [9]. Curing the polymer (e.g., polyurethane acrylate resin) in the etched Si master yields slanted structures with the designed angle and aspect ratio (Figure 1D). Taken together with the ultraviolet (UV)-assisted capillary force lithography, the etched Si master can be further used to create two-level hierarchical PUA hairs for enhanced robustness to a rough surface (<20 μm). In addition to the hierarchical structure [9,68,69], the shape of the tip was also found to have an influential role on adhesion strength [70] (e.g., mushroom-like and spatula tips [50,64,71,72] were shown to have higher adhesion than flat and round tips [73]).

On a separate route to structural design for enhanced adhesion, octopus suckers that reversibly adhere to wettable surfaces provided another source of inspiration [74–77]. The strong adhesion in both dry and wet environments results from the lower pressure in the octopus suckers than that of the environment. Using an external control (e.g., suctioning system [78], vacuum pump [79], dielectric elastomer actuator [80], or magnetic actuated film [81]), the biomimetic system can be easily created. The miniaturization of the system was also achieved through the use of lithographic processes [82,83]. In one attempt to create nanoscale suction cups [84], a non-close-packed self-assembled silica nanoparticle array served as an etching mask to prepare mushroom-like structures consisting of polymer stems and silica caps (Figure 2A (i)). Drying and peeling a polyvinyl alcohol (PVA) film from the etched structure created a replica with embedded silica nanoparticles (Figure 2A (ii)), which formed a mold to yield silicone polymer with nano-sucker structures (Figure 2A (iii)). By controlling the meniscus of a liquid precursor from applied pressure, a simple molding process from the mold with different surface energies could yield artificial micro-suckers with well-controlled cross-sectional profiles [85]. The adhesion in both dry and underwater environments was shown to increase as the curvature of the cross-sectional profile increased, due to the increased contact area from the preload (Figure 2B). In the wet environment, a model that combined the suction effect and capillary interaction [86] captured the experimental observation. When the elastomeric PDMS film with suction-cup structures was covered by thermoresponsive hydrogel of poly(N-isopropylacrylamide) (pNIPAM), a resulting smart adhesive pad could respond to temperature change, with an increased temperature inducing an increased volume and decreased pressure in the suction cup due to the deformation of the pNIPAM layer (Figure 2C) [87].

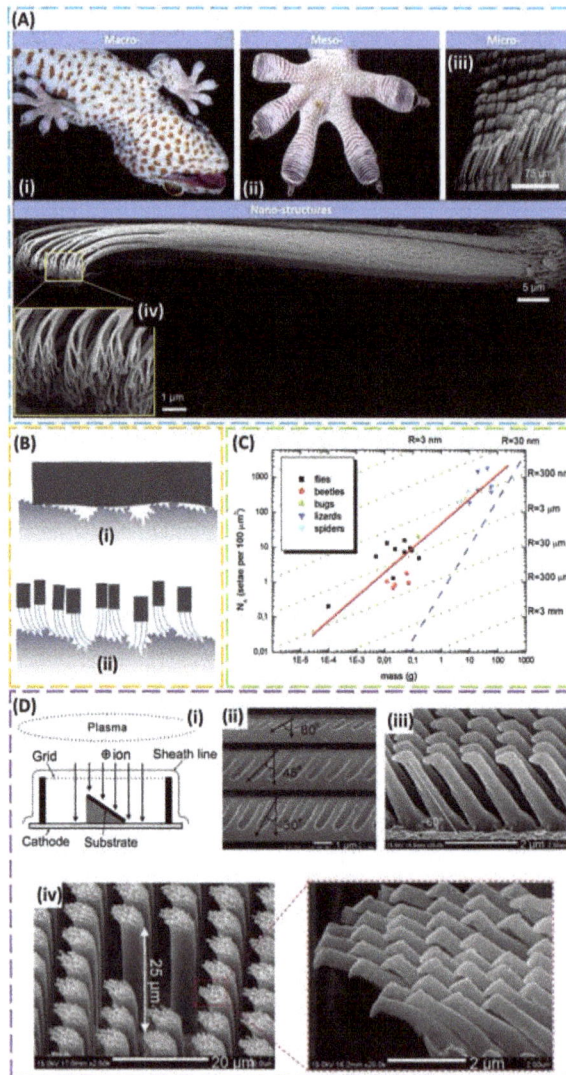

Figure 1. Gecko-inspired dry adhesive. (**A**) (**i**) A close-up view of a gecko climbing a glass wall and (**ii**) mesostructure of gecko toe pad with (**iii**) microscale array of high-aspect-ratio setae structure. (**iv**) Cryo-SEM view of a single seta (ca. 110 μm in length and 4.2 μm in diameter) that branches at the tips into 100–1000 more structures known as spatulas. Reproduced with permission from Reference [56]; Copyright 2006, The Company of Biologists. (**B**) Schematic illustrations of adhesives with (**i**) a relatively flat surface and (**ii**) an array of fibrils in contact with a rough surface. Reproduced with permission from Reference [49]; Copyright 2010, John Wiley and Sons. (**C**) Dependence of the terminal element density (N_A) of the attachment pads on the body mass (m) in hairy-pad systems of diverse animal groups. The red line that fits all data corresponds to the self-similarity criterion with a slope of ~0.67. The green lines (slope of ~0.33) correspond to the model with a curvature invariance of contacts with radius R. The blue line shows the approximate limit of maximum contact for such attachment devices. Reproduced with permission from Reference [47]; Copyright 2003, United States National Academy of Sciences. (**D**) (**i**) A schematic showing the mechanism of a plasma sheath with a Faraday cage, where

ions are incident on the substrate surface in a direction normal to the grid plane for an angled etching in the silicon substrate. (ii) SEM images of polySi etch profiles with angles of 30°/45°/60°, and (iii) polyurethane acrylate (PUA) nano-hairs formed on a poly(ethylene terephthalate) (PET) film substrate with a slanted angle of 60°. (iv) SEM image of two-level hierarchical PUA hairs with well-defined high aspect ratio over a large area using two-step ultraviolet (UV)-assisted capillary force lithography; the inset shows angled nano-hairs on the tip. Reproduced with permission from Reference [9]; Copyright 2009, United States National Academy of Sciences.

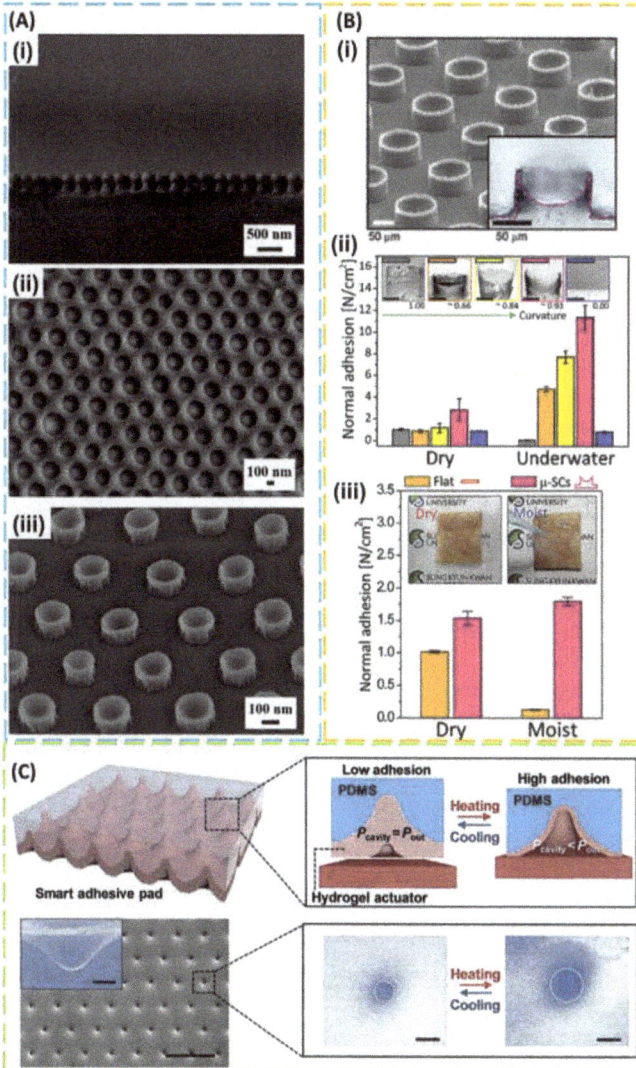

Figure 2. Octopus-inspired adhesive. (**A**) SEM images of (**i**) mushroom-like structures consisting of silica caps and polymer stems on a silicon wafer prepared by plasma etching; (**ii**) a polyvinyl alcohol (PVA) composite film embedding silica nanoparticles; (**iii**) a polydimethylsiloxane (PDMS) nano-sucker array obtained from the PVA composite mold. Reproduced with permission from Reference [84]; Copyright

2017, American Chemical Society. (**B**) (**i**) SEM image and cross-sectional optical image of micro-suckers (100 μm in diameter and 75 μm in height); (**ii**) dry/wet adhesion strength increases as the curvature of the cross-sectional profile increases; (**iii**) pull-off strength for PDMS-based micro-suckers and a non-patterned PDMS patch against a rough pigskin, with dry and moist pigskins shown in the inset. Reproduced with permission from Reference [85]; Copyright 2018, John Wiley and Sons. (**C**) Schematic representation and SEM image (scale bar: 10 μm) of micro-cavity arrays within a smart adhesive pad and its corresponding switchable adhesion mechanism (i.e., temperature-dependent hydrogel actuation in response to the environmental temperature). The inset in the bottom right shows SEM images of a single hole on the smart adhesive pad at a low temperature (left, T ≈ 3 °C) and a high temperature (right, T ≈ 40 °C) (scale bar: 500 nm). Reproduced with permission from Reference [85]; Copyright 2018, John Wiley and Sons.

3. Design of the Material for Use in Wet Conditions

Though mechanical properties (e.g., the previously discussed structural designs and Young's modulus of the structure [88–90]) showed significant effects on the strength of adhesion in the dry environment, many of them are compromised in the wet environment. In order to address the challenge, several bio-inspired materials [91–93] and their integration with structural designs [7,94] were explored. As a celebrated biological model for wet adhesion [95], mussels were shown to attach virtually all types of inorganic and organic surfaces, including classically adhesion-resistant materials such as poly(tetrafluoroethylene) (PTFE). Clues to this versatility may lie in the amino-acid composition of the specialized adhesive proteins that contain the catecholic amino acid 3,4-dihydroxy-L-phenylalanine (DOPA) and lysine [96]. DOPA and other catechol components perform well as adhesives. With inspiration from both geckos and mussels, a flexible organic nano-adhesive "geckel" was created by dip-coating the gecko-foot-mimetic PDMS pillar array in an ethanol solution of mussel-adhesive-protein-mimetic polymer (Figure 3A (i)) [7]. With a high catechol content, the adhesive monomer, dopamine methacrylamide (DMA), was used in a free-radical polymerization to synthesize poly(dopamine methacrylamide-co-methoxyethyl acrylate) (p(DMA-co-MEA)) as the mussel-adhesive-protein-mimetic polymer. The addition of a p(DMA-co-MEA) coating on the pillars enhanced the wet adhesion by nearly 15 times (Figure 3A (ii)), and this geckel nanoadhesive maintained its adhesive performance for over 1000 contact cycles in both dry and wet environments (Figure 3A (iii)).

Containing both catechol (DOPA) and amine (lysine) functional groups, dopamine as a simple-molecule compound also shows promise to achieve adhesion to a wide spectrum of materials [97]. The adherent polydopamine (PDA) coating produced by self-polymerization of dopamine can also serve as a versatile platform to graft various organic molecules and biomacromolecules for secondary surface-mediated reactions (Figure 3B). Taken together with its biocompatibility and hydrophilicity, polydopamine-based materials demonstrated great potential toward biomedical applications, ranging from cell adhesion/encapsulating/patterning to tissue engineering and re-endothelialization of vascular devices [98,99]. As a versatile building block, PDA was also integrated with other materials such as a hydrogel. However, hydrogel is often associated with long-term instability from water evaporation and physical changes from use at relatively extreme temperatures [100,101]. In order to provide hydrogel with long-term stable operation in a wide temperature window, a glycerol–water (GW) mixture as the binary solvent was used in hydrogel development (Figure 3C) [102], as glycerol is a well-known nontoxic anti-freezing agent. Incorporating PDA-decorated carbon nanotubes (CNTs) as conductive nano-fillers into hydrogel imparts good conductivity (~8 S/m), enhanced toughness (~2000 J/m^2), and excellent adhesion (57 kPa to porcine skin) to the resulting GW hydrogel. Due to its advantages of good adhesiveness and anti-heating or anti-freezing properties, GW hydrogel demonstrated its capability to protect skin from damage in harsh environments (e.g., during frostbite or burn) by serving as an excellent wearable dressing. Other challenges of hydrogel also include foreign body response and poor mechanical properties (i.e., toughness and stretchability). The former could be attenuated by encapsulating mesenchymal stem cells within the hydrogel, such as poly(ethylene glycol) (PEC) [103], and the latter is addressed by the design of tough hydrogel [104–106], discussed in Section 4.

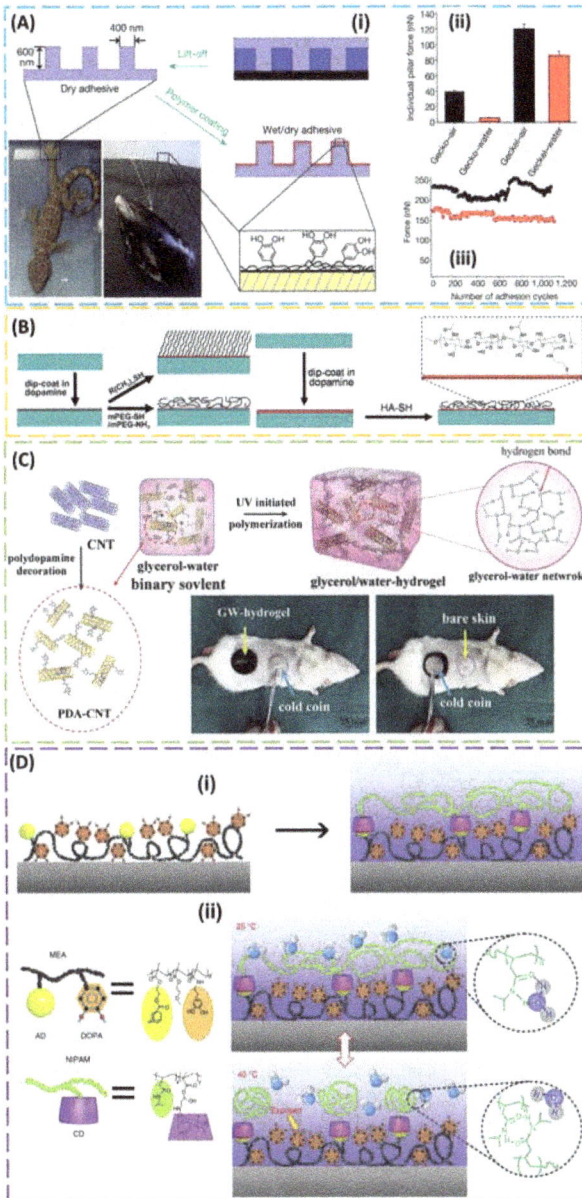

Figure 3. Mussel-inspired wet adhesives. (**A**) (**i**) PDMS casting onto the master is followed by curing, and lift-off results in gecko-foot-mimetic nano-pillar arrays. Next, the fabricated nanopillars are coated with a mussel-adhesive-protein-mimetic polymer that contains catechols, a key component of wet adhesive proteins found in mussel holdfasts. (**ii**) Adhesion force per pillar for gecko and geckel in water (red)

and air (black). (iii) Performance of geckel adhesive during multiple contact cycles in water (red) and air (black) with error bars to represent the standard deviation. Reproduced with permission from Reference [7]; Copyright 2007, Nature Publishing Group. (**B**) Schematic illustrations of alkanethiol monolayer and poly(ethylene glycol) (PEG) polymer grafting, as well as the glycosaminoglycan hyaluronic acid (HA) conjugation to polydopamine (PDA)-coated surfaces. Reproduced with permission from Reference [7]; Copyright 2007, American Association for the Advancement of Science. (**C**) Polydopamine decorating carbon nanotubes (PDA-CNTs) in water results in PDA-CNT dispersion. Copolymerizing acrylamide (AM) and acrylic acid (AA) monomers in the PDA-CNT dispersion forms a glycerol–water (GW) hydrogel in a glycerol–water binary solvent. Frostbite and burn models on rats' back skin demonstrate anti-freezing and anti-burning properties of the adhesive GW-hydrogel. Reproduced with permission from Reference [102]; Copyright 2018, John Wiley and Sons. (**D**) (i) Schematic diagram showing the specific procedure to prepare the wet adhesive, where dip-coating an adhesive guest copolymer poly(3,4-dihydroxy-L-phenylalanine) (pDOPA)/adamantine (AD)/methoxyethyl acrylate (MEA) on a clean Si substrate is followed by the self-assembly of host copolymer poly(N-isopropylacrylamide) (pNIPAM)/cyclodextrin (CD) using host–guest molecular recognition. (ii) Schematic drawing showing the tunable wet adhesion that responds to a local temperature trigger. When the local temperature of the adhesive is below lower critical solution temperature (LCST), the pNIPAM easily forms intermolecular hydrogen bonding with adjacent water molecules, and the infused water layer transforms the pNIPAM side chains to a swelling layer, which spatially stabilizes and confines the underlying adhesive moiety, DOPA. On the other hand, heating above the LCST leads to a phase transition and the collapse of pNIPAM-CD chains to form numerous agglomerates, exposing the adhesive group. Reproduced with permission from Reference [10]; Copyright 2017, Nature Publishing Group.

In order to provide reversible and tunable wet adhesion that responds to a local temperature trigger in an on-demand manner, a mussel-inspired guest-adhesive copolymer was combined with a thermoresponsive host copolymer [10]. The guest copolymer pDOPA/adamantine (AD)/methoxyethyl acrylate (MEA) consists of a mussel-inspired adhesive DOPA polymer, a guest motif adamantine (AD), and a methoxyethyl acrylate (MEA) monomer as a hydrophobic matrix to enhance the wet adhesion of DOPA (Figure 3D (i)). In the host copolymer pNIPAM/cyclodextrin (CD), the poly(N-isopropylacrylamide) (pNIPAM) undergoes a reversible lower critical solution temperature (LCST) phase transition from a swollen hydrated state to a shrunken dehydrated state when heated above the LCST and β-cyclodextrin (β-CD) is the host molecule providing selective binding with the AD moiety in the guest copolymer. Dip-coating the as-prepared guest copolymer on the target substrate surface (e.g., Si, Ti, Al, glass, PTFE, or PDMS) allows the self-assembly of the host copolymer through the host–guest interaction. When the local temperature of the adhesive is below the LCST, the swollen hydrated pNIPAM spatially confines and stabilizes the underneath adhesive moiety, DOPA, through the host–guest interaction, resulting in a dramatically screened interaction area and reduced adhesion. In contrast, the collapsed pNIPAM exposes the adhesive moiety, DOPA, when the local temperature is above LCST (Figure 3D (ii)). The versatile demonstration of the wet adhesive also goes from inorganic (Si, Ti, Al, and glass) to organic surfaces (PDMS and PTFE). In addition, the gecko-like surface structure (e.g., an array of PDMS posts with a diameter of 5 μm and and height of 10 μm), discussed in Section 2, was explored to further enhance the interfacial adhesion strength, which is in direct contrast with the gecko-like dry adhesive.

Although mussel-like wet adhesion was successfully realized, typical catechol functionalization and solution processing entail complex components and steps. In order to reduce the complexity, synthetic low-molecular-weight catecholic zwitterionic surfactants were developed to adhere to diverse surfaces with very strong adhesion (~50 mJ/m^2) [107]. Based on catechol-modified amphiphilic poly(propylene oxide)/poly(ethylene oxide) (PPO-PEO) block copolymers, a mechanically tough zero- or negative-swelling mussel-inspired surgical adhesive was synthesized, minimizing the weakening mechanism from swelling [108]. The range of zero to −25 % swelling was achieved through a hydrophobic collapse of PPO blocks upon heating to physiological temperature. The lap shear adhesion measurements of decellularized porcine dermis show nearly 50 kPa adhesive strength. Although the single-layer mussel-like adhesion is effective, a layer-by-layer (LbL) assembly may be explored to further enhance the adhesion strength due to the versatile control in the assembly process (e.g., introducing sodium chloride in the assembly process yields an adhesion enhanced by two orders of magnitude [109]).

4. Adhesion to Biological Tissues

When it comes to adhesion to biological tissues such as skin, several additional challenges are encountered, including soft properties, multiscale roughness, and biocompatibility. For instance, adhesives based on chemical bonding may irritate the skin and cause discomfort upon removal due to strong adhesion. Though several commercial adhesives were used in bio-integrated electronics on the skin [110,111], their applications are limited by their given properties, and the adhesion strength was also shown to be dependent on the target tissue. Taking a synthetic tissue adhesive (i.e., Dermabond®, 2-octyl cyanoacrylate) as an example, its adhesion to collagen films was observed to be 40 times that when compared with its adhesion to muscle tissue, due to increased wetting (and the decreased contact angle) of the Dermabond® adhesive on the collagen film [12]. In the two classes of tissue adhesives, biologic (e.g., fibrin glue) and synthetic (e.g., *n*-butyl-2-cyanoacrylate) [112], a variety of different bonding mechanisms were explored (e.g., physical interaction, mechanical interlocking, and chemical bonding) [113,114]. As an extensively used synthetic polymer for tissue engineering, polyethylene glycol (PEG) was used with chondroitin sulfate (CS) to form a biodegradable CS–PEG adhesive hydrogel that can covalently bond to proteins in tissue or to collagen in the extracellular matrix via amide bonds, improving the adhesion strength by ten times that of fibrin glue (Figure 4A) [115]. In a separate effort, a buckypaper (BP) film produced from oxidized multi-walled carbon nanotubes demonstrated enhanced adhesion to the rimmed muscular fascia of the abdominal wall of New Zealand female rabbit during both peeling and shearing tests, due to soft tissue deformation from water suction resulting in water bridge formation and BP–tissue mechanical interlocking, respectively (Figure 4B) [116].

Because of the self-cleaning effect and adaptation capability, the gecko-inspired fibrillar adhesive demonstrated repeatable and restorable adhesion to the skin surface over multiple cycles of use [5]. Though the adhesion does not show a direct correlation with classical roughness parameters, strong adhesion was shown to decrease significantly when surface roughness increased [117–119] and a clear correlation was even observed when a newly integrated roughness parameter was introduced [120]. In order to address these challenges, special considerations have to be given to the design of the adhesion layer. In one effort to utilize the unique advantages of the gecko-inspired fibrillar adhesive, a composite adhesive was designed by coating polymer microfibers with a skin interfacing material (e.g., vinylsiloxane) to form mushroom-shaped tips (Figure 4C) [26]. As a member of the family of silicone rubbers, the biocompatible vinylsiloxane (VS) approved for biomedical applications (e.g., forming dental impression) can fully cross-link and form covalent bonds with the PDMS microfibers within a few minutes at room temperature, enabling its direct cross-linking on and conformal contact to the skin surface with multiscale roughness. Due to the high flexibility and strong attachment to the skin surface, the adhesive layer shows an efficient strain signal to the strain sensor integrated on top, thereby significantly increasing the signal-to-noise ratio when compared with medical tape or fibrillar adhesive film fully immersed into a flat vs. film. The adhesion strength of the composite adhesive to a wet skin surface was also shown to be comparable to that of a dry environment. In contrast to the gecko-inspired adhesive, the octopus-inspired adhesive that relies on the pressure difference was less affected by the surface roughness. Such an adhesive also showed comparable adhesion strength to pigskin in moist conditions (40% of area covered with droplets) to that in dry conditions, even when hairs were present [85]. When integrated with physiological sensors and drug-delivery actuators, the octopus-inspired adhesive allowed sensitive biometric measurements and transdermal drug delivery through tight skin coupling (Figure 4D) [83].

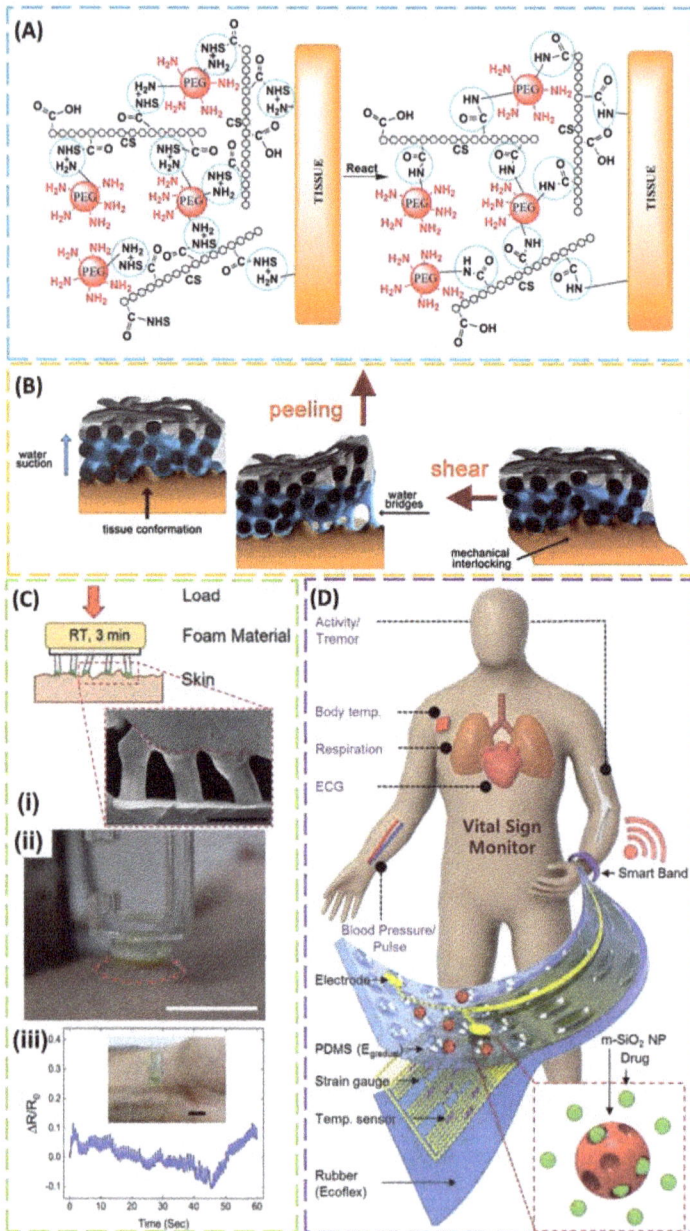

Figure 4. Adhesives on soft tissues. (**A**) Chondroitin sulfate *N*-succinimidyl succinate (CS–NHS) reacts with primary amines of both six-arm polyethylene glycol amine PEG–(NH$_2$)$_6$ and proteins of the tissue to form a covalently bonded hydrogel to the tissue. Reproduced with permission from Reference [115]; Copyright 2010, Elsevier. (**B**) When the buckypaper (BP) comes into contact with the wet biological tissue, water suction leads to deformation of the soft tissue and conformal contact of the BP to the tissue.

The resulting water bridge formation and BP–tissue mechanical interlocking help enhance the adhesion during peeling and shearing tests. Reproduced with permission from Reference [116]; Copyright 2013, American Chemical Society. (**C**) (**i**) Inking vinylsiloxane (VS) precursor onto uniformly shaped cylindrical microfibers followed by curing vs. directly onto the skin surface. The inset shows the cross-sectional SEM image of an adhesive film attached to an artificial skin replica, where the red dashed line indicates the interface between the skin-adhesive film and artificial skin. (**ii**) Optical image of a skin-adhesive film attached to the human forearm during the retraction cycle of the adhesion experiment (dashed red line indicates the interfacing border between the adhesive film and the skin; scale bar: 1 cm). (**iii**) The output signal from a microfibrillar strain sensor mounted onto the radial artery of the wrist; the inset shows a photograph of the strain sensor attached to the wrist through the inking and printing process. Reproduced with permission from Reference [26]; Copyright 2017, John Wiley and Sons. (**D**) Schematic illustration of the multifunctional electronic patch integrated onto an octopus-inspired adhesive layer with miniaturized suction cups (mSCs). The multifunctional electronic patch is capable of monitoring vital signs from electrodes, strain gauges, and temperature sensors, as well as transdermally delivering drugs loaded in mesoporous silica nanoparticles via iontophoresis. The smart band (connected to the mSC patch) is used to provide wireless functionalities and a power source. Reproduced with permission from Reference [83]; Copyright 2015, John Wiley and Sons.

Relying on relatively weak physical interactions, existing tissue adhesives (including mussel-inspired adhesives) are associated with low adhesive energy on the order of 10 J/m^2 [121], which is far from ideal, especially when compared with the example in nature. For instance, cartilage bonds to bones with an adhesion energy of 800 J/m^2 [122]. In order to achieve high adhesion, a synergy from an adhesive surface layer and a dissipative matrix was explored (Figure 5A) [105]. The design was inspired by a sticky and tough secretion from slug *Arion subfuscus* [123], which may arise from two interpenetrating networks of polymers [124]. The strong adhesion from the adhesive surface layer to the tissue substrate can be achieved through electrostatic interactions, covalent bonds, or physical interpenetration. Meanwhile, the energy dissipation through hysteresis in the matrix amplifies the effective adhesion energy. As the surface of tissues or cells is negatively charged, a bridging polymer that bears positively charged primary amine groups enables covalent binding via electrostatic attraction. In the case of a permeable target surface, the bridging polymer penetrating into the target forms a physical entanglement and a chemical anchor for the adhesive. As for the dissipative matrix, a substrate that can dissipate energy is used. By exploiting the synergy of these two factors, a class of tough adhesives demonstrated high adhesion energy (~1000 J/m^2) on wet surfaces. In vivo demonstrations included strong adhesion to a beating porcine heart in the presence of blood, heart sealants to prevent liquid leakage, and hemostatic dressing for a deep wound (Figure 5B). This simple yet effective strategy opens up a wide range of applications, including tissue adhesives, wound dressing, and tissue repair.

In order to provide a reversible adhesion that can respond to the external stimuli, a responsive polymer was explored to diffuse into and form an entangled network with two polymer networks of two wet materials for enhanced adhesion upon the trigger of one signal (e.g., in one pH range), while being soluble and separating both wet materials upon trigger removal (e.g., in the other pH range) (Figure 5C) [125]. In the demonstration, several stitching polymers were identified to cover the full range of pH (e.g., cellulose forms a network for pKa < 13, alginate for pKa < 3.5, chitosan for pKa > 6.5, and poly(4-aminostyrene) for pKa > 4.5). Adhesion energy as high as 1000 J/m^2 could be achieved when the stitching polymer introduced hysteresis in the wet hydrogel materials. The demonstrated strong adhesion also went beyond the hydrogel to various porcine tissues (e.g., skin, liver, heart, artery, and stomach) and the skin was shown to exhibit a relatively high adhesion energy (100 J/m^2) due to its relatively high toughness.

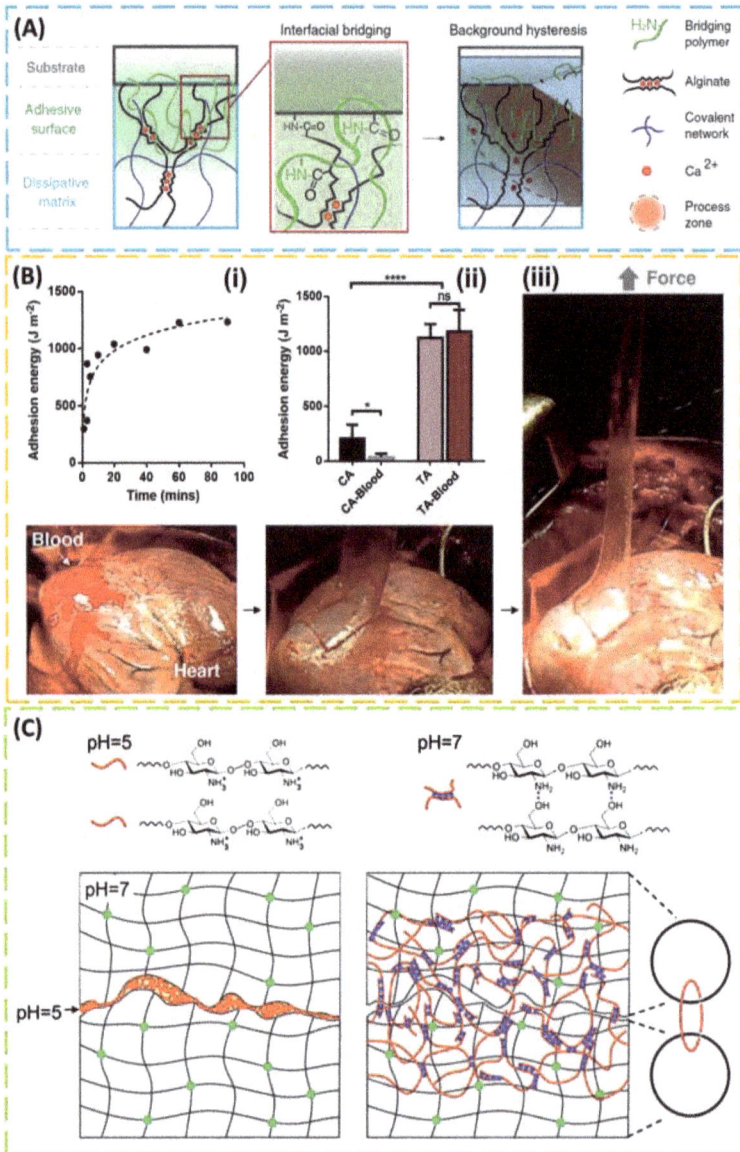

Figure 5. Tough and topological adhesives. (**A**) Tough adhesives (TAs) consist of a dissipative matrix (light-blue square), made of a hydrogel containing both ionically (calcium; red circles) cross-linked and covalently cross-linked polymers (black and blue lines), and an adhesive surface that contains a bridging polymer with primary amines (green lines). The bridging polymer can penetrate into the TA and the substrate (light-green region) to facilitate covalent-bond formation. In the presence of a crack, the process zone (orange area) dissipates significant amounts of energy as ionic bonds between alginate chains and calcium ions break. (**B**) (**i**) Tough adhesives exhibit a rapid increase in adhesion energy to porcine skin over time. (**ii**) Comparing with cyanoacrylate (CA), TAs showed strong adhesion even when the porcine skin was exposed to blood in the in vitro experiment. n = 4–6. (**iii**) The TAs were

further tested in an in vivo experiment on a beating porcine heart with blood exposure. (**A,B**) Reproduced with permission from References [105]; Copyright 2017, American Association for the Advancement of Science. (**C**) Chitosan chains dissolve in water at pH 5 and form a network in water at pH 7. Placing an aqueous solution of chitosan of pH 5 between two hydrogels (or biological tissues) of pH 7 is followed by the diffusion of chitosan chains into the two hydrogels, forming a network that topologically entangles with the networks of both hydrogels. Reproduced with permission from Reference [125]; Copyright 2018, John Wiley and Sons.

5. Conclusions and Future Perspectives

In order to robustly adhere bio-integrated devices to soft biological tissues, an adhesive layer with tunable adhesion is of great interest. The capability to switch between strong and weak adhesion would allow the use of strong adhesion to efficiently transfer vital signals to the device for accurate measurement, followed by the use of weak adhesion for easy removal. However, this tunable adhesion to the tissue surface has long been a challenge due to multiscale roughness, wet conditions, biocompatibility, and natural motion, among many others. Thanks to the recent developments that shed light on the underlying mechanisms of the remarkable adhesion observed in several animal species, great strides were made, and effective strategies ranging from structural design for dry adhesion to novel material synthesis for wet conditions were explored to yield adhesives that can match or even outperform those from nature. The importance of the developed adhesives also goes beyond bio-integrated devices to cell culture [126,127] and to tissue glues that can potentially replace sutures in clinical practice. When combined with the tunable properties of the adhesive, tissue glues would promise repeatable use, which can dramatically reduce the cost and pave the way for commercialization. Despite great strides made in the field of tissue adhesives, several challenges still exist, including fabricating high-aspect-ratio fibrillar structures with diameters down to submicron scales [40], long-term reliability of the tissue adhesives to diverse wet surfaces, tunable properties in the adhesives to accommodate dynamic changes in target tissues, and integration with multifunctional electronics for real-time sensing and closed-loop control [128,129]. In the burgeoning field of tissue adhesives, different testing methods and tissue models are used to evaluate the adhesive properties of newly developed structures and materials. Thus, it is a bit challenging to directly compare the results reported by different research groups. It would be desirable to have standardized testing procedures and tissue models in place to allow for direct comparison among the newly developed tissue adhesives. Nevertheless, the challenges simply represent a small fraction of the great opportunities for future development, which may require the collective wisdom of material scientists, chemists, mechanical engineers, and clinicians, among many others.

Author Contributions: Z.Y. and H.C. wrote the manuscript and assembled the figures. H.C. led the preparation of the manuscript and contributed to editorial modifications of the overall text.

Funding: This research was funded by the start-up fund provided by the Engineering Science and Mechanics Department, College of Engineering, and Materials Research Institute at The Pennsylvania State University. The authors also acknowledge the support from the ASME Haythornthwaite Foundation Research Initiation Grant, the Doctoral New Investigator grant from the American Chemical Society Petroleum Research Fund, and the Dorothy Quiggle Career Development Professorship in Engineering and Global Engineering Leadership Program at Penn State.

Conflicts of Interest: The authors declare no competing financial interests.

References

1. Shull, K.R. Contact mechanics and the adhesion of soft solids. *Mater. Sci. Eng. R Rep.* **2002**, *36*, 1–45. [CrossRef]
2. Zhu, J.; Dexheimer, M.; Cheng, H. Reconfigurable systems for multifunctional electronics. *npj Flex. Electron.* **2017**, *1*, 8. [CrossRef]
3. Choi, S.; Lee, H.; Ghaffari, R.; Hyeon, T.; Kim, D.H. Recent Advances in Flexible and Stretchable Bio-Electronic Devices Integrated with Nanomaterials. *Adv. Mater.* **2016**, *28*, 4203–4218. [CrossRef] [PubMed]

4.   Rogers, J.A.; Someya, T.; Huang, Y. Materials and mechanics for stretchable electronics. *Science* **2010**, *327*, 1603–1607. [CrossRef] [PubMed]
5.   Kwak, M.K.; Jeong, H.E.; Suh, K.Y. Rational design and enhanced biocompatibility of a dry adhesive medical skin patch. *Adv. Mater.* **2011**, *23*, 3949–3953. [CrossRef] [PubMed]
6.   Poulard, C.; Restagno, F.; Weil, R.; Léger, L. Mechanical tuning of adhesion through micro-patterning of elastic surfaces. *Soft Matter* **2011**, *7*, 2543–2551. [CrossRef]
7.   Lee, H.; Lee, B.P.; Messersmith, P.B. A reversible wet/dry adhesive inspired by mussels and geckos. *Nature* **2007**, *448*, 338–341. [CrossRef] [PubMed]
8.   Boesel, L.F.; Cremer, C.; Arzt, E.; Del Campo, A.; Greiner, C.; Arzt, E.; del Campo, A. Gecko-inspired surfaces: A path to strong and reversible dry adhesives. *Adv. Mater.* **2010**, *22*, 2125–2137. [CrossRef] [PubMed]
9.   Jeong, H.E.; Lee, J.-K.; Kim, H.N.; Moon, S.H.; Suh, K.Y. A nontransferring dry adhesive with hierarchical polymer nanohairs. *Proc. Natl. Acad. Sci. USA* **2009**, *106*, 5639–5644. [CrossRef] [PubMed]
10.   Zhao, Y.; Wu, Y.; Wang, L.; Zhang, M.; Chen, X.; Liu, M.; Fan, J.; Liu, J.; Zhou, F.; Wang, Z. Bio-inspired reversible underwater adhesive. *Nat. Commun.* **2017**, *8*, 2218. [CrossRef] [PubMed]
11.   Campolo, D.; Jones, S.; Fearing, R.S. Fabrication of gecko foot-hair like nano structures and adhesion to random rough surfaces. In Proceedings of the 2003 Third IEEE Conference on Nanotechnology, San Francisco, CA, USA, 12–14 August 2003.
12.   Bochyńska, A.I.; Hannink, G.; Buma, P.; Grijpma, D.W. Adhesion of tissue glues to different biological substrates. *Polym. Adv. Technol.* **2017**, *28*, 1294–1298. [CrossRef]
13.   Bruns, T.B.; Worthington, J.M. Using tissue adhesive for wound repair: A practical guide to Dermabond. *Am. Fam. Physician* **2000**, *61*, 1383–1388. [PubMed]
14.   Dermabond, Ò.; Corneal, S.; Leung, G.Y.S.; Peponis, V.; Varnell, E.D.; Lam, D.S.C.; Kaufman, H.E. Preliminary In Vitro Evaluation of 2-Octyl Cyanoacrylate to seal corneal incisions. *Cornea* **2005**, *24*, 998–999.
15.   Bae, W.G.; Kim, D.; Kwak, M.K.; Ha, L.; Kang, S.M.; Suh, K.Y. Enhanced Skin Adhesive Patch with Modulus-Tunable Composite Micropillars. *Adv. Healthc. Mater.* **2013**, *2*, 109–113. [CrossRef] [PubMed]
16.   Gelinck, G.H.; Huitema, H.E.A.; Van Veenendaal, E.; Cantatore, E.; Schrijnemakers, L.; Van Der Putten, J.B.P.H.; Geuns, T.C.T.; Beenhakkers, M.; Giesbers, J.B.; Huisman, B.H.; et al. Flexible active-matrix displays and shift registers based on solution-processed organic transistors. *Nat. Mater.* **2004**, *3*, 106–110. [CrossRef] [PubMed]
17.   Kim, R.H.; Bae, M.H.; Kim, D.G.; Cheng, H.; Kim, B.H.; Kim, D.H.; Li, M.; Wu, J.; Du, F.; Kim, H.S.; et al. Stretchable, transparent graphene interconnects for arrays of microscale inorganic light emitting diodes on rubber substrates. *Nano Lett.* **2011**, *11*, 3881–3886. [CrossRef] [PubMed]
18.   Sekitani, T.; Nakajima, H.; Maeda, H.; Fukushima, T.; Aida, T.; Hata, K.; Someya, T. Stretchable active-matrix organic light-emitting diode display using printable elastic conductors. *Nat. Mater.* **2009**, *8*, 494–499. [CrossRef] [PubMed]
19.   Liao, X.; Liao, Q.; Yan, X.; Liang, Q.; Si, H.; Li, M.; Wu, H.; Cao, S.; Zhang, Y. Flexible and highly sensitive strain sensors fabricated by pencil drawn for wearable monitor. *Adv. Funct. Mater.* **2015**, *25*, 2395–2401. [CrossRef]
20.   Trung, T.Q.; Lee, N.E. Flexible and Stretchable Physical Sensor Integrated Platforms for Wearable Human-Activity Monitoringand Personal Healthcare. *Adv. Mater.* **2016**, *28*, 4338–4372. [CrossRef] [PubMed]
21.   Jang, K.I.; Han, S.Y.; Xu, S.; Mathewson, K.E.; Zhang, Y.; Jeong, J.W.; Kim, G.T.; Webb, R.C.; Lee, J.W.; Dawidczyk, T.J.; et al. Rugged and breathable forms of stretchable electronics with adherent composite substrates for transcutaneous monitoring. *Nat. Commun.* **2014**, *5*, 4779. [CrossRef] [PubMed]
22.   Webb, R.C.; Bonifas, A.P.; Behnaz, A.; Zhang, Y.; Yu, K.J.; Cheng, H.; Shi, M.; Bian, Z.; Liu, Z.; Kim, Y.S.; et al. Ultrathin conformal devices for precise and continuous thermal characterization of human skin. *Nat. Mater.* **2013**, *12*, 938–944. [CrossRef] [PubMed]
23.   Xu, L.; Gutbrod, S.R.; Bonifas, A.P.; Su, Y.; Sulkin, M.S.; Lu, N.; Chung, H.J.; Jang, K.I.; Liu, Z.; Ying, M.; et al. 3D multifunctional integumentary membranes for spatiotemporal cardiac measurements and stimulation across the entire epicardium. *Nat. Commun.* **2014**, *5*, 3329. [CrossRef] [PubMed]
24.   Dewire, J.; Calkins, H. State-of-the-art and emerging technologies for atrial fibrillation ablation. *Nat. Rev. Cardiol.* **2010**, *7*, 129–138. [CrossRef] [PubMed]
25.   Son, D.; Lee, J.; Qiao, S.; Ghaffari, R.; Kim, J.; Lee, J.E.; Song, C.; Kim, S.J.; Lee, D.J.; Jun, S.W.; et al. Multifunctional wearable devices for diagnosis and therapy of movement disorders (Support Information). *Nat. Nanotechnol.* **2014**, *9*, 397–404. [CrossRef] [PubMed]

26. Drotlef, D.M.; Amjadi, M.; Yunusa, M.; Sitti, M. Bioinspired Composite Microfibers for Skin Adhesion and Signal Amplification of Wearable Sensors. *Adv. Mater.* **2017**, *29*, 1–8. [CrossRef] [PubMed]

27. Carlson, A.; Bowen, A.M.; Huang, Y.; Nuzzo, R.G.; Rogers, J.A. Transfer printing techniques for materials assembly and micro/nanodevice fabrication. *Adv. Mater.* **2012**, *24*, 5284–5318. [CrossRef] [PubMed]

28. Yang, S.Y.; Carlson, A.; Cheng, H.; Yu, Q.; Ahmed, N.; Wu, J.; Kim, S.; Sitti, M.; Ferreira, P.M.; Huang, Y.; et al. Elastomer surfaces with directionally dependent adhesion strength and their use in transfer printing with continuous roll-to-roll applications. *Adv. Mater.* **2012**, *24*, 2117–2122. [CrossRef] [PubMed]

29. Kim, S.; Carlson, A.; Cheng, H.; Lee, S.; Park, J.K.; Huang, Y.; Rogers, J.A. Enhanced adhesion with pedestal-shaped elastomeric stamps for transfer printing. *Appl. Phys. Lett.* **2012**, *100*, 171909. [CrossRef]

30. Carlson, A.; Kim-Lee, H.J.; Wu, J.; Elvikis, P.; Cheng, H.; Kovalsky, A.; Elgan, S.; Yu, Q.; Ferreira, P.M.; Huang, Y.; et al. Shear-enhanced adhesiveless transfer printing for use in deterministic materials assembly. *Appl. Phys. Lett.* **2011**, *98*, 264104. [CrossRef]

31. Gao, Y.; Cheng, H. Assembly of Heterogeneous Materials for Biology and Electronics: From Bio-Inspiration to Bio-Integration. *J. Electron. Packag.* **2017**, *139*, 020801. [CrossRef]

32. Wang, T.; Ramnarayanan, A.; Cheng, H. Real time analysis of bioanalytes in healthcare, food, zoology and botany. *Sensors* **2018**, *18*, 5. [CrossRef] [PubMed]

33. Cheng, H.; Yi, N. Dissolvable tattoo sensors: From science fiction to a viable technology. *Phys. Scr.* **2017**, *92*, 13001. [CrossRef]

34. Stoppa, M.; Chiolerio, A. Wearable electronics and smart textiles: A critical review. *Sensors* **2014**, *14*, 11957–11992. [CrossRef] [PubMed]

35. Hwang, S.W.; Tao, H.; Kim, D.H.; Cheng, H.; Song, J.K.; Rill, E.; Brenckle, M.A.; Panilaitis, B.; Won, S.M.; Kim, Y.S.; et al. A physically transient form of silicon electronics. *Science* **2012**, *337*, 1640–1644. [CrossRef] [PubMed]

36. Hwang, S.W.; Park, G.; Cheng, H.; Song, J.K.; Kang, S.K.; Yin, L.; Kim, J.H.; Omenetto, F.G.; Huang, Y.; Lee, K.M.; et al. 25th anniversary article: Materials for high-performance biodegradable semiconductor devices. *Adv. Mater.* **2014**, *26*, 1992–2000. [CrossRef] [PubMed]

37. Hwang, S.W.; Song, J.K.; Huang, X.; Cheng, H.; Kang, S.K.; Kim, B.H.; Kim, J.H.; Yu, S.; Huang, Y.; Rogers, J.A. High-performance biodegradable/transient electronics on biodegradable polymers. *Adv. Mater.* **2014**, *26*, 3905–3911. [CrossRef] [PubMed]

38. Cheng, H. Inorganic dissolvable electronics: Materials and devices for biomedicine and environment. *J. Mater. Res.* **2016**, *31*, 2549–2570. [CrossRef]

39. Cheng, H.; Vepachedu, V. Recent development of transient electronics. *Theor. Appl. Mech. Lett.* **2016**, *6*, 21–31. [CrossRef]

40. Eisenhaure, J.; Kim, S. A review of the state of dry adhesives: Biomimetic structures and the alternative designs they inspire. *Micromachines* **2017**, *8*, 125. [CrossRef]

41. Li, Y.; Krahn, J.; Menon, C. Bioinspired Dry Adhesive Materials and Their Application in Robotics: A Review. *J. Bionic Eng.* **2016**, *13*, 181–199. [CrossRef]

42. Zhou, M.; Pesika, N.; Zeng, H.; Tian, Y.; Israelachvili, J. Recent advances in gecko adhesion and friction mechanisms and development of gecko-inspired dry adhesive surfaces. *Friction* **2013**, *1*, 114–129. [CrossRef]

43. Autumn, K.; Sitti, M.; Liang, Y.A.; Peattie, A.M.; Hansen, W.R.; Sponberg, S.; Kenny, T.W.; Fearing, R.; Israelachvili, J.N.; Full, R.J. Evidence for van der Waals adhesion in gecko setae. *Proc. Natl. Acad. Sci. USA* **2002**, *99*, 12252–12256. [CrossRef] [PubMed]

44. Gao, H.; Yao, H. Shape insensitive optimal adhesion of nanoscale fibrillar structures. *Proc. Natl. Acad. Sci. USA* **2004**, *101*, 7851–7856. [CrossRef] [PubMed]

45. Varenberg, M.; Pugno, N.M.; Gorb, S.N. Spatulate structures in biological fibrillar adhesion. *Soft Matter* **2010**, *6*, 3269–3272. [CrossRef]

46. Jagota, A. Mechanics of Adhesion through a Fibrillar Microstructure. *Integr. Comp. Biol.* **2002**, *42*, 1140–1145. [CrossRef] [PubMed]

47. Arzt, E.; Gorb, S.; Spolenak, R. From micro to nano contacts in biological attachment devices. *Proc. Natl. Acad. Sci. USA* **2003**, *100*, 10603–10606. [CrossRef] [PubMed]

48. Aksak, B.; Murphy, M.P.; Sitti, M. Gecko inspired micro-fibrillar adhesives for wall climbing robots on micro/nanoscale rough surfaces. In Proceedings of the 2008 IEEE International Conference on Robotics and Automation, Pasadena, CA, USA, 19–23 May 2008.

49. Kamperman, M.; Kroner, E.; del Campo, A.; McMeeking, R.M.; Arzt, E. Functional Adhesive Surfaces with "Gecko" Effect: The Concept of Contact Splitting. *Adv. Eng. Mater.* **2010**, *12*, 335–348. [CrossRef]

50. Gorb, S.; Varenberg, M.; Peressadko, A.; Tuma, J. Biomimetic mushroom-shaped fibrillar adhesive microstructure. *J. R. Soc. Interface* **2007**, *4*, 271–275. [CrossRef] [PubMed]

51. Lee, J.; Fearing, R.S. Contact self-cleaning of synthetic gecko adhesive from polymer microfibers. *Langmuir* **2008**, *24*, 10587–10591. [CrossRef] [PubMed]

52. Hansen, W.R.; Autumn, K. Evidence for self-cleaning in gecko setae. *Proc. Natl. Acad. Sci. USA* **2005**, *102*, 385–389. [CrossRef] [PubMed]

53. Shahsavan, H.; Zhao, B. Biologically inspired enhancement of pressure-sensitive adhesives using a thin film-terminated fibrillar interface. *Soft Matter* **2012**, *8*, 8281–8284. [CrossRef]

54. Yao, H.; Rocca, G.D.; Guduru, P.R.; Gao, H. Adhesion and sliding response of a biologically inspired fibrillar surface: Experimental observations. *J. R. Soc. Interface* **2008**, *5*, 723–733. [CrossRef] [PubMed]

55. Lundberg, D. Flow Conditioners. *Control Eng.* **2006**, *53*. [CrossRef]

56. Autumn, K.; Majidi, C.; Groff, R.E.; Dittmore, A.; Fearing, R. Effective elastic modulus of isolated gecko setal arrays. *J. Exp. Biol.* **2006**, *209*, 3558–3568. [CrossRef] [PubMed]

57. Geim, A.K.; Dubonos, S.V.; Grigorieva, I.V.; Novoselov, K.S.; Zhukov, A.A.; Shapoval, S.Y. Microfabricated adhesive mimicking gecko foot-hair. *Nat. Mater.* **2003**, *2*, 461–463. [CrossRef] [PubMed]

58. Mahdavi, A.; Ferreira, L.; Sundback, C.; Nichol, J.W.; Chan, E.P.; Carter, D.J.D.; Bettinger, C.J.; Patanavanich, S.; Chignozha, L.; Ben-Joseph, E.; et al. A biodegradable and biocompatible gecko-inspired tissue adhesive. *Proc. Natl. Acad. Sci. USA* **2008**, *105*, 2307–2312. [CrossRef] [PubMed]

59. Majidi, C.; Groff, R.E.; Maeno, Y.; Schubert, B.; Baek, S.; Bush, B.; Maboudian, R.; Gravish, N.; Wilkinson, M.; Autumn, K.; et al. High friction from a stiff polymer using microfiber arrays. *Phys. Rev. Lett.* **2006**, *97*, 076103. [CrossRef] [PubMed]

60. Jeong, H.E.; Lee, S.H.; Kim, P.; Suh, K.Y. Stretched polymer nanohairs by nanodrawing. *Nano Lett.* **2006**, *6*, 1508–1513. [CrossRef] [PubMed]

61. Ge, L.; Sethi, S.; Ci, L.; Ajayan, P.M.; Dhinojwala, A. Carbon nanotube-based synthetic gecko tapes. *Proc. Natl. Acad. Sci. USA* **2007**, *104*, 10792–10795. [CrossRef] [PubMed]

62. Qu, L.; Dai, L.; Stone, M.; Xia, Z.; Wang, Z.L. Carbon nanotube arrays with strong shear binding-on and easy normal lifting-off. *Science* **2008**, *322*, 238–242. [CrossRef] [PubMed]

63. Aksak, B.; Murphy, M.P.; Sitti, M. Adhesion of biologically inspired vertical and angled polymer microfiber arrays. *Langmuir* **2007**, *23*, 3322–3332. [CrossRef] [PubMed]

64. Murphy, M.P.; Aksak, B.; Sitti, M. Gecko-inspired directional and controllable adhesion. *Small* **2009**, *5*, 170–175. [CrossRef] [PubMed]

65. Reddy, S.; Arzt, E.; Del Campo, A. Bioinspired surfaces with switchable adhesion. *Adv. Mater.* **2007**, *19*, 3833–3837. [CrossRef]

66. Kim, T.I.; Jeong, H.E.; Suh, K.Y.; Lee, H.H. Stooped nanohairs: Geometry-controllable, unidirectional, reversible, and robust Gecko-like dry adhesive. *Adv. Mater.* **2009**, *21*, 2276–2281. [CrossRef]

67. Röhrig, M.; Thiel, M.; Worgull, M.; Hölscher, H. 3D Direct laser writing of nano- and microstructured hierarchical gecko-mimicking surfaces. *Small* **2012**, *8*, 3009–3015. [CrossRef] [PubMed]

68. Murphy, M.P.; Kim, S.; Sitti, M. Enhanced adhesion by gecko-inspired hierarchical fibrillar adhesives. *ACS Appl. Mater. Interfaces* **2009**, *1*, 849–855. [CrossRef] [PubMed]

69. Greiner, C.; Arzt, E.; Del Campo, A. Hierarchical gecko-like adhesives. *Adv. Mater.* **2009**, *21*, 479–482. [CrossRef]

70. Del Campo, A.; Greiner, C.; Arzt, E. Contact shape controls adhesion of bioinspired fibrillar surfaces. *Langmuir* **2007**, *23*, 10235–10243. [CrossRef] [PubMed]

71. Kim, S.; Sitti, M. Biologically inspired polymer microfibers with spatulate tips as repeatable fibrillar adhesives. *Appl. Phys. Lett.* **2006**, *89*, 261911. [CrossRef]

72. Sameoto, D.; Menon, C. A low-cost, high-yield fabrication method for producing optimized biomimetic dry adhesives. *J. Micromech. Microeng.* **2009**, *19*, 115002. [CrossRef]

73. Mengüç, Y.; Yang, S.Y.; Kim, S.; Rogers, J.A.; Sitti, M. Gecko-inspired controllable adhesive structures applied to micromanipulation. *Adv. Funct. Mater.* **2012**, *22*, 1246–1254. [CrossRef]

74. Smith, A.M. Negative Pressure Generated By Octopus Suckers: A Study of the Tensile Strength of Water in Nature. *J. Exp. Biol.* **1991**, *157*, 257–271.

75. Tramacere, F.; Beccai, L.; Kuba, M.; Gozzi, A.; Bifone, A.; Mazzolai, B. The Morphology and Adhesion Mechanism of Octopus vulgaris Suckers. *PLoS ONE* **2013**, *8*, e65074. [CrossRef] [PubMed]
76. Kier, W.M. The Structure and Adhesive Mechanism of Octopus Suckers. *Integr. Comp. Biol.* **2002**, *42*, 1146–1153. [CrossRef] [PubMed]
77. Tramacere, F.; Pugno, N.M.; Kuba, M.J.; Mazzolai, B. Unveiling the morphology of the acetabulum in octopus suckers and its role in attachment. *Interface Focus* **2014**, *5*, 1–5. [CrossRef] [PubMed]
78. Tramacere, F.; Beccai, L.; Mattioli, F.; Sinibaldi, E.; Mazzolai, B. Artificial adhesion mechanisms inspired by octopus suckers. In Proceedings of the IEEE International Conference on Robotics and Automation, Saint Paul, MN, USA, 14–18 May 2012; pp. 3846–3851.
79. Tomokazu, T.; Kikuchi, S.; Suzuki, M.; Aoyagi, S. Vacuum gripper imitated octopus sucker-effect of liquid membrane for absorption. In Proceedings of the IEEE International Conference on Intelligent Robots and Systems, Hamburg, Germany, 28 September–2 October 2015; pp. 2929–2936.
80. Follador, M.; Tramacere, F.; Mazzolai, B. Dielectric elastomer actuators for octopus inspired suction cups. *Bioinspir. Biomim.* **2014**, *9*. [CrossRef] [PubMed]
81. Yu, Q.; Chen, F.; Zhou, H.; Yu, X.; Cheng, H.; Wu, H. Design and Analysis of Magnetic-Assisted Transfer Printing. *J. Appl. Mech.* **2018**, *85*, 101009. [CrossRef]
82. Chang, W.Y.; Wu, Y.; Chung, Y.C. Facile fabrication of ordered nanostructures from protruding nanoballs to recessional nanosuckers via solvent treatment on covered nanosphere assembled monolayers. *Nano Lett.* **2014**, *14*, 1546–1550. [CrossRef] [PubMed]
83. Choi, M.K.; Park, O.K.; Choi, C.; Qiao, S.; Ghaffari, R.; Kim, J.; Lee, D.J.; Kim, M.; Hyun, W.; Kim, S.J.; et al. Cephalopod-Inspired Miniaturized Suction Cups for Smart Medical Skin. *Adv. Healthc. Mater.* **2016**, *5*, 80–87. [CrossRef] [PubMed]
84. Chen, Y.C.; Yang, H. Octopus-Inspired Assembly of Nanosucker Arrays for Dry/Wet Adhesion. *ACS Nano* **2017**, *11*, 5332–5338. [CrossRef] [PubMed]
85. Baik, S.; Kim, J.; Lee, H.J.; Lee, T.H.; Pang, C. Highly Adaptable and Biocompatible Octopus-Like Adhesive Patches with Meniscus-Controlled Unfoldable 3D Microtips for Underwater Surface and Hairy Skin. *Adv. Sci.* **2018**, *5*. [CrossRef] [PubMed]
86. Li, K.; Cai, S. Wet adhesion between two soft layers. *Soft Matter* **2014**, *10*, 8202–8209. [CrossRef] [PubMed]
87. Lee, H.; Um, D.S.; Lee, Y.; Lim, S.; Kim, H.-j.; Ko, H. Octopus-Inspired Smart Adhesive Pads for Transfer Printing of Semiconducting Nanomembranes. *Adv. Mater.* **2016**, *28*, 7457–7465. [CrossRef] [PubMed]
88. Matuda, N.; Baba, S.; Kinbara, A. Internal stress, young's modulus and adhesion energy of carbon films on glass substrates. *Thin Solid Films* **1981**, *81*, 301–305. [CrossRef]
89. Schneider, A.; Francius, G.; Obeid, R.; Schwinté, P.; Hemmerlé, J.; Frisch, B.; Schaaf, P.; Voegel, J.-C.; Senger, B.; Picart, C. Polyelectrolyte Multilayers with a Tunable Young's Modulus: Influence of Film Stiffness on Cell Adhesion. *Langmuir* **2006**, *22*, 1193–1200. [CrossRef] [PubMed]
90. Pan, T.; Pharr, M.; Ma, Y.; Ning, R.; Yan, Z.; Xu, R.; Feng, X.; Huang, Y.; Rogers, J.A. Experimental and Theoretical Studies of Serpentine Interconnects on Ultrathin Elastomers for Stretchable Electronics. *Adv. Funct. Mater.* **2017**, *27*, 1702589. [CrossRef]
91. Pena-Francesch, A.; Akgun, B.; Miserez, A.; Zhu, W.; Gao, H.; Demirel, M.C. Pressure Sensitive Adhesion of an Elastomeric Protein Complex Extracted From Squid Ring Teeth. *Adv. Funct. Mater.* **2014**, *24*, 6227–6233. [CrossRef]
92. Zhao, Q.; Woog Lee, D.; Kollbe Ahn, B.; Seo, S.; Kaufman, Y.; Israelachvili, J.N.; Herbert Waite, J. Underwater contact adhesion and microarchitecture in polyelectrolyte complexes actuated by solvent exchange. *Nat. Mater.* **2016**, *15*, 407–412. [CrossRef] [PubMed]
93. Ahn, B.K. Perspectives on Mussel-Inspired Wet Adhesion. *J. Am. Chem. Soc.* **2017**, *139*, 10166–10171. [CrossRef] [PubMed]
94. Rao, P.; Sun, T.L.; Chen, L.; Takahashi, R.; Shinohara, G.; Guo, H.; King, D.R.; Kurokawa, T.; Gong, J.P. Tough Hydrogels with Fast, Strong, and Reversible Underwater Adhesion Based on a Multiscale Design. *Adv. Mater.* **2018**, *30*, 1801884. [CrossRef] [PubMed]
95. Waite, J.H.; Tanzer, M.L. Polyphenolic substance of Mytilus edulis: Novel adhesive containing L-dopa and hydroxyproline. *Science* **1981**, *212*, 1038–1040. [CrossRef] [PubMed]
96. Waite, J.H.; Qin, X. Polyphosphoprotein from the adhesive pads of *Mytilus edulis*. *Biochemistry* **2001**, *40*, 2887–2893. [CrossRef] [PubMed]

97. Lee, H.; Dellatore, S.M.; Miller, W.M.; Messersmith, P.B. Mussel-inspired surface chemistry for multifunctional coatings. *Science* **2007**, *318*, 426–430. [CrossRef] [PubMed]
98. Liu, Y.; Ai, K.; Lu, L. Polydopamine and its derivative materials: Synthesis and promising applications in energy, environmental, and biomedical fields. *Chem. Rev.* **2014**, *114*, 5057–5115. [CrossRef] [PubMed]
99. Lynge, M.E.; Van Der Westen, R.; Postma, A.; Städler, B. Polydopamine—A nature-inspired polymer coating for biomedical science. *Nanoscale* **2011**, *3*, 4916–4928. [CrossRef] [PubMed]
100. Lee, Y.-Y.; Kang, H.-Y.; Gwon, S.H.; Choi, G.M.; Lim, S.-M.; Sun, J.-Y.; Joo, Y.-C. A Strain-Insensitive Stretchable Electronic Conductor: PEDOT:PSS/Acrylamide Organogels. *Adv. Mater.* **2016**, *28*, 1636–1643. [CrossRef] [PubMed]
101. Yuk, H.; Zhang, T.; Parada, G.A.; Liu, X.; Zhao, X. Skin-inspired hydrogel–elastomer hybrids with robust interfaces and functional microstructures. *Nat. Commun.* **2016**, *7*, 12028. [CrossRef] [PubMed]
102. Han, L.; Liu, K.; Wang, M.; Wang, K.; Fang, L.; Chen, H.; Zhou, J.; Lu, X. Mussel-Inspired Adhesive and Conductive Hydrogel with Long-Lasting Moisture and Extreme Temperature Tolerance. *Adv. Funct. Mater.* **2018**, *28*, 1704195. [CrossRef]
103. Swartzlander, M.D.; Blakney, A.K.; Amer, L.D.; Hankenson, K.D.; Kyriakides, T.R.; Bryant, S.J. Immunomodulation by mesenchymal stem cells combats the foreign body response to cell-laden synthetic hydrogels. *Biomaterials* **2015**, *41*, 79–88. [CrossRef] [PubMed]
104. Han, L.; Lu, X.; Liu, K.; Wang, K.; Fang, L.; Weng, L.-T.; Zhang, H.; Tang, Y.; Ren, F.; Zhao, C.; et al. Mussel-Inspired Adhesive and Tough Hydrogel Based on Nanoclay Confined Dopamine Polymerization. *ACS Nano* **2017**, *11*, 2561–2574. [CrossRef] [PubMed]
105. Li, J.; Celiz, A.D.; Yang, J.; Yang, Q.; Wamala, I.; Whyte, W.; Seo, B.R.; Vasilyev, N.V.; Vlassak, J.J.; Suo, Z.; et al. Tough adhesives for diverse wet surfaces. *Science* **2017**, *357*, 378–381. [CrossRef] [PubMed]
106. Zhao, P.; Wei, K.; Feng, Q.; Chen, H.; Wong, D.S.H.; Chen, X.; Wu, C.-C.; Bian, L. Mussel-mimetic hydrogels with defined cross-linkers achieved via controlled catechol dimerization exhibiting tough adhesion for wet biological tissues. *Chem. Commun.* **2017**, *53*, 12000–12003. [CrossRef] [PubMed]
107. Ahn, B.K.; Das, S.; Linstadt, R.; Kaufman, Y.; Martinez-Rodriguez, N.R.; Mirshafian, R.; Kesselman, E.; Talmon, Y.; Lipshutz, B.H.; Israelachvili, J.N.; et al. High-performance mussel-inspired adhesives of reduced complexity. *Nat. Commun.* **2015**, *6*, 8663. [CrossRef] [PubMed]
108. Barrett, D.G.; Bushnell, G.G.; Messersmith, P.B. Mechanically Robust, Negative-Swelling, Mussel-Inspired Tissue Adhesives. *Adv. Healthc. Mater.* **2013**, *2*, 745–755. [CrossRef] [PubMed]
109. Pettersson, T.; Pendergraph, S.A.; Utsel, S.; Marais, A.; Gustafsson, E.; Wågberg, L. Robust and tailored wet adhesion in biopolymer thin films. *Biomacromolecules* **2014**, *15*, 4420–4428. [CrossRef] [PubMed]
110. Koh, A.; Kang, D.; Xue, Y.; Lee, S.; Pielak, R.M.; Kim, J.; Hwang, T.; Min, S.; Banks, A.; Bastien, P.; et al. A soft, wearable microfluidic device for the capture, storage, and colorimetric sensing of sweat. *Sci. Transl. Med.* **2016**, *8*, 366ra165. [CrossRef] [PubMed]
111. Wang, S.; Xu, J.; Wang, W.; Wang, G.-J.N.; Rastak, R.; Molina-Lopez, F.; Chung, J.W.; Niu, S.; Feig, V.R.; Lopez, J.; et al. Skin electronics from scalable fabrication of an intrinsically stretchable transistor array. *Nature* **2018**, *555*, 83–88. [CrossRef] [PubMed]
112. Panda, A.; Kumar, S.; Kumar, A.; Bansal, R.; Bhartiya, S. Fibrin glue in ophthalmology. *Indian J. Ophthalmol.* **2009**, *57*, 371–379. [CrossRef] [PubMed]
113. Khurana, A.; Parker, S.; Goel, V.; Alderman, P.M. Dermabond wound closure in primary hip arthroplasty. *Acta Orthop. Belg.* **2008**, *74*, 349–353. [PubMed]
114. Agarwal, A.; Kumar, D.A.; Jacob, S.; Baid, C.; Agarwal, A.; Srinivasan, S. Fibrin glue-assisted sutureless posterior chamber intraocular lens implantation in eyes with deficient posterior capsules. *J. Cataract Refract. Surg.* **2008**, *34*, 1433–1438. [CrossRef] [PubMed]
115. Strehin, I.; Nahas, Z.; Arora, K.; Nguyen, T.; Elisseeff, J. A versatile pH sensitive chondroitin sulfate-PEG tissue adhesive and hydrogel. *Biomaterials* **2010**, *31*, 2788–2797. [CrossRef] [PubMed]
116. Martinelli, A.; Carru, G.A.; D'Ilario, L.; Caprioli, F.; Chiaretti, M.; Crisante, F.; Francolini, I.; Piozzi, A. Wet adhesion of buckypaper produced from oxidized multiwalled carbon nanotubes on soft animal tissue. *ACS Appl. Mater. Interfaces* **2013**, *5*, 4340–4349. [CrossRef] [PubMed]
117. Barreau, V.; Hensel, R.; Guimard, N.K.; Ghatak, A.; McMeeking, R.M.; Arzt, E. Fibrillar Elastomeric Micropatterns Create Tunable Adhesion Even to Rough Surfaces. *Adv. Funct. Mater.* **2016**, *26*, 4687–4694. [CrossRef]

118. Bauer, C.T.; Kroner, E.; Fleck, N.A.; Arzt, E. Hierarchical macroscopic fibrillar adhesives: In situ study of buckling and adhesion mechanisms on wavy substrates. *Bioinspir. Biomim.* **2015**, *10*. [CrossRef] [PubMed]
119. Stark, A.Y.; Palecek, A.M.; Argenbright, C.W.; Bernard, C.; Brennan, A.B.; Niewiarowski, P.H.; Dhinojwala, A. Gecko Adhesion on Wet and Dry Patterned Substrates. *PLoS ONE* **2015**, *10*, e0145756. [CrossRef] [PubMed]
120. Kasem, H.; Varenberg, M. Effect of counterface roughness on adhesion of mushroom-shaped microstructure. *J. R. Soc. Interface* **2013**, *10*, 20130620. [CrossRef] [PubMed]
121. Dastjerdi, A.K.; Pagano, M.; Kaartinen, M.T.; McKee, M.D.; Barthelat, F. Cohesive behavior of soft biological adhesives: Experiments and modeling. *Acta Biomater.* **2012**, *8*, 3349–3359. [CrossRef] [PubMed]
122. Moretti, M.; Wendt, D.; Schaefer, D.; Jakob, M.; Hunziker, E.B.; Heberer, M.; Martin, I. Structural characterization and reliable biomechanical assessment of integrative cartilage repair. *J. Biomech.* **2005**, *38*, 1846–1854. [CrossRef] [PubMed]
123. Pawlicki, J.M. The effect of molluscan glue proteins on gel mechanics. *J. Exp. Biol.* **2004**, *207*, 1127–1135. [CrossRef] [PubMed]
124. Wilks, A.M.; Rabice, S.R.; Garbacz, H.S.; Harro, C.C.; Smith, A.M. Double-network gels and the toughness of terrestrial slug glue. *J. Exp. Biol.* **2015**, *218*, 3128–3137. [CrossRef] [PubMed]
125. Yang, J.; Bai, R.; Suo, Z. Topological Adhesion of Wet Materials. *Adv. Mater.* **2018**, *30*, 1–7. [CrossRef] [PubMed]
126. Théry, M.; Pépin, A.; Dressaire, E.; Chen, Y.; Bornens, M. Cell distribution of stress fibres in response to the geometry of the adhesive environment. *Cell Motil. Cytoskelet.* **2006**, *63*, 341–355. [CrossRef] [PubMed]
127. Chen, X.; Cortez-Jugo, C.; Choi, G.H.; Björnmalm, M.; Dai, Y.; Yoo, P.J.; Caruso, F. Patterned Poly(dopamine) Films for Enhanced Cell Adhesion. *Bioconjug. Chem.* **2017**, *28*, 75–80. [CrossRef] [PubMed]
128. Malki, M.; Fleischer, S.; Shapira, A.; Dvir, T. Gold Nanorod-Based Engineered Cardiac Patch for Suture-Free Engraftment by Near IR. *Nano Lett.* **2018**, *18*, 4069–4073. [CrossRef] [PubMed]
129. Liang, S.; Zhang, Y.; Wang, H.; Xu, Z.; Chen, J.; Bao, R.; Tan, B.; Cui, Y.; Fan, G.; Wang, W.; et al. Paintable and Rapidly Bondable Conductive Hydrogels as Therapeutic Cardiac Patches. *Adv. Mater.* **2018**, *30*, 1704235. [CrossRef] [PubMed]

micromachines

MDPI

Review

Novel Nano-Materials and Nano-Fabrication Techniques for Flexible Electronic Systems

Kyowon Kang [†], Younguk Cho [†] and Ki Jun Yu *

School of Electrical Engineering, Yonsei University, Seoul 03722, Korea; kyowon.kang@yonsei.ac.kr (K.K.); hot9198@yonsei.ac.kr (Y.C.)
* Correspondence: kijunyu@yonsei.ac.kr; Tel.: +82-221-232-769
† These authors contributed equally to this work.

Received: 30 April 2018; Accepted: 24 May 2018; Published: 28 May 2018

Abstract: Recent progress in fabricating flexible electronics has been significantly developed because of the increased interest in flexible electronics, which can be applied to enormous fields, not only conventional in electronic devices, but also in bio/eco-electronic devices. Flexible electronics can be applied to a wide range of fields, such as flexible displays, flexible power storages, flexible solar cells, wearable electronics, and healthcare monitoring devices. Recently, flexible electronics have been attached to the skin and have even been implanted into the human body for monitoring biosignals and for treatment purposes. To improve the electrical and mechanical properties of flexible electronics, nanoscale fabrications using novel nanomaterials are required. Advancements in nanoscale fabrication methods allow the construction of active materials that can be combined with ultrathin soft substrates to form flexible electronics with high performances and reliability. In this review, a wide range of flexible electronic applications via nanoscale fabrication methods, classified as either top-down or bottom-up approaches, including conventional photolithography, soft lithography, nanoimprint lithography, growth, assembly, and chemical vapor deposition (CVD), are introduced, with specific fabrication processes and results. Here, our aim is to introduce recent progress on the various fabrication methods for flexible electronics, based on novel nanomaterials, using application examples of fundamental device components for electronics and applications in healthcare systems.

Keywords: flexible electronics; nano-fabrication; top-down approaches; bottom-up approaches

---

1. Introduction

Currently, standard electronic devices require multi-functional platforms in a limited substrate area. To keep in pace with the demands of the device users, nanofabrication processing for flexible electronics [1–5] has been deeply studied in a variety of fields, such as multifunctional energy storage devices [6–8], wearable devices for human healthcare [9–12], and displays [13–15], with great practicality and reliability.

More specifically, the approaches to device fabrication with mechanical flexibility can be divided into two dominant processes, namely, top-down [16–19] and bottom-up [20–22] approaches. The main differences between the top-down and the bottom-up approaches are the mechanisms and methods for the initial state, such as from the bulk-structure to the desired flexible device for top-down, or the complete opposite method for bottom-up. Here, we introduce ultrathin nano-material structures and fabrication methods that allow assemblies of heterogeneously integrated functional materials onto soft substrates, with all of the active components maintaining excellent electronic functionality. In addition, we focus on well-designed unconventional biomedical devices using the top-down approach and multifunctional flexible sensors from the bottom-up approach, including a flexible graphene transistor, photonic device, $MoSe_2$ transistor, and light emitting diode (LED) for the initial components. These flexible and stretchable electronic systems, with performances that reach or exceed the levels of conventional electronic systems, are classified. This review summarizes some recent progress in the field of micro- and nano-electronics and showcases the practical applications of the novel electronic devices.

2. Novel Devices Designed by Top-Down Nanofabrication

2.1. Introduction to the Top-Down Approach

The top-down approach is a process where the structure is removed from a larger substance. In particular, in the area of macro-electronics, the top-down approach is more desirable because various materials can be produced by conventional lithographic processes [23,24] and etching techniques [25–27]. The lateral dimensions of this method can range from tens of nanometers to the millimeter scale. Additionally, freestanding single-crystal sheet types can be created from wafers by employing an embedded release layer so as to yield flexible systems. Thus, nanostructures are synthesized by etching the layers on a substrate. However, it is impossible to conduct the entire procedure on a flexible substrate, because certain processes, such as the doping process and chemical vapor deposition (CVD), require high temperatures that significantly deform flexible substrates [28]. To resolve this issue, a transfer-printing process and a method of releasing from the rigid substrate are integrated to form flexible electronics [29–36]. Here, various flexible electronic devices that are fabricated using the top-down approach, are introduced.

2.2. Transfer-Printed Graphene Lines for Flexible Transistor

Recent work demonstrates that graphene-based electrolyte-gated transistors (EGTs) can be introduced as a platform via transfer printing with a silicon stencil, to produce graphene lines [37]. The transfer printing process, which is conducted with a viscous graphene-flake ink, enables transistors to maintain their great electrical performance, while achieving a high printing resolution with flexibility and a scalability at low cost; such results are not easily achieved by the conventional rigid substrate-based fabricating processes. Figure 1a shows the schematics of the detailed fabrication steps, starting with the spin-casting of a Cytop film on the micro-lithographically patterned Si wafer to design a mold. Then, graphene ink is squeezed through the line-holes of the Cytop/Si stencil, forming five groups of graphene lines using the differences in surface energies and wetting properties at the interface of the mold and ink. After the annealing of the graphene film, the graphene lines are ready to be transferred onto a flexible substrate. Figure 1b shows the specific procedures for transfer printing graphene lines onto a polyethylene terephthalate (PET) substrate. A liquid Norland Optical Adhesive (NOA73, Norland Products, Inc., Cranbury, NJ, USA) is coated on the graphene lines, and then the ultraviolet (UV) light is illuminated at the backside of the O_2-plasma-treated PET substrate. The graphene lines are subsequently peeled-off from the Cytop/Si mold by the UV-cured NOA73 adhesive layer. Figure 1c shows the flexible aerosol-jet printed transistors, which are fabricated on the basis of poly(3-hexylthiophene) (P3HT, Sigma-Aldrich, Saint Louis, MO, USA) as the semiconductor, ion gel for the gate dielectric and graphene lines for the source-drain contact pads. Figure 1d,e present the transfer characteristics and output characteristics of the device, respectively. These two graphs show the operation of the transistor with negligible drain current (I_D) hysteresis and clear I_D saturation. This unconventional fabrication process of graphene transistors provides benefits for circuit design, and by increasing the packing density with transfer printing, assures compatibility for the scalable manufacturing of the flexible devices.

❖ **Top-down approaches**

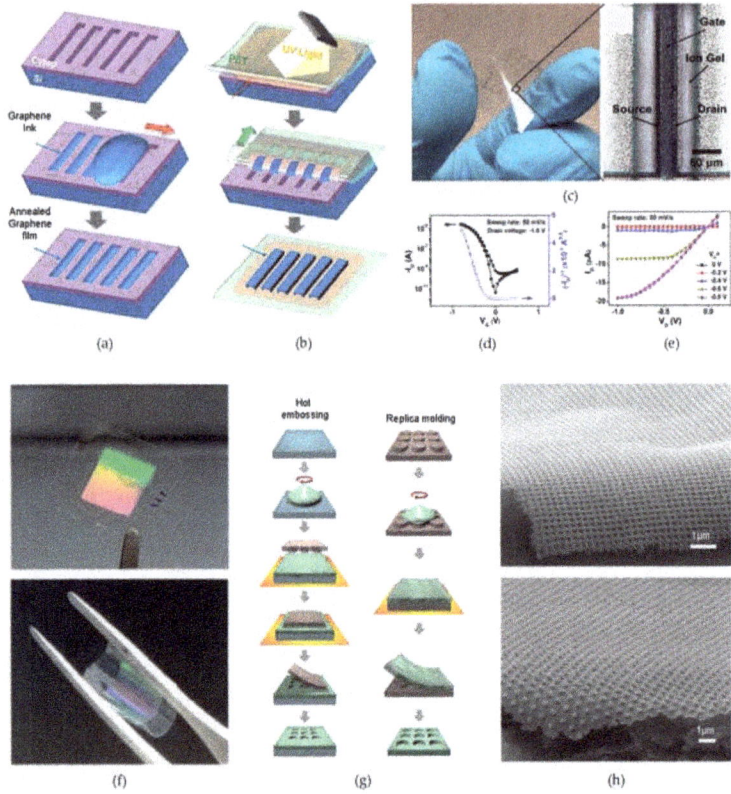

(a) (b) (c) (d) (e)

(f) (g) (h)

Figure 1. Graphene lines-based flexible transistor/hydroxypropyl cellulose (HPC) photonic thin film with a hexagonal nanopillar structure: (**a**) a schematic of the fabrication of the Cytop/Si mold; (**b**) simple procedures for transferring graphene lines from the Cytop/Si mold to the polyethylene terephthalate (PET) flexible substrate; (**c**) an image of flexible electrolyte-gated transistor (EGT) arrays, the magnified picture shows the composition of each transistor; (**d,e**) the transfer and output characteristics of the graphene electrodes, Reproduced with permission from 2017 ACS NANO [37]; (**f**) images of an HPC photonic crystal (top) and its mechanical flexibility with a free-standing property of the design (bottom); (**g**) schematics of two fabrication processes to make HPC photonic films-hot embossing and replica molding methods. The blue-colored slide indicates the glass substrate, the green one for HPC, and the brown one for hard polydimethylsiloxane (h-PDMS); (**h**) special hexagonal nanopillar images obtained by scanning electron microscopy (SEM) in the lateral view. Paper substrates are used to imprint the predesigned nanopattern. Reproduced with permission from 2018 Nature [38].

2.3. Flexible Photonic Device with Hexagonal Structures

Flexible photonic and plasmonic devices based on eco-friendly hydroxypropyl cellulose (HPC) are introduced in Figure 1f [38]. In general, when the nanoscale celluloses exist in the form of fibers, they may not be seen as white in color, but are seen as possessing a transparent or iridescent property. This optical property can be arbitrarily adjusted by the structure of the particles via either the amorphous state or the single-crystalline state. Using the special chiral behavior [39–42] that is possessed by an HPC solution, a simple nanostructured design provides an enhanced photoluminescence and the potential of a HPC plasmonic membrane in the role of an optical device.

The core fabrication process is focused on exploiting soft lithography [43–47] rather than using conventional photolithography or a growth method. Soft lithography realizes a cost-effective process with large-area patterning, ensuring enhanced reproducibility. Figure 1g shows two specific approaches that are used to fabricate photonic HPC films. The first one is the hot-embossing process. Spin-casted HPC on a rigid substrate is heated, which is followed by a hard polydimethylsiloxane (h-PDMS) compound-based patterning. Then, the HPC photonic film is transferred onto the flexible substrate in order to complete the device fabrication. The second method is to use a replica molding process. On the surface of an h-PDMS mold, the HPC solution is poured and dried by applying heat. Finally, a free-standing, flexible photonic HPC film is obtained by peeling from the PDMS mold. The complete structures that are acquired from both methods are nearly identical, but the hot-embossing process presents a product with a better optical property. After thdepositing and patterning a metal on the HPC photonic film, the fabrication of the HPC film can be completed. Figure 1h shows the scanning electron microscopy (SEM) image of the plasmonic HPC film that has been patterned with arrays of a hexagonal lattice structure. This unique nanostructure upgrades the photoluminescence by maintaining its iridescent property with amplified optical extinction spectra.

2.4. Novel Biomedical Electronics—Piezoelectric Probes for Biopsy Diagnosis

Flexible electronics that are fabricated by top-down approaches can be directly applied to medical applications because of the absence of a mechanical mismatch between the flexible electronics and the organs or tissues of the human body. The contents that are to be introduced next are new bio-integrated devices, which can be widely applied in medical treatment/diagnosis. Here, we introduce the devices that can be formed by lifting off a thin layer from a temporary substrate using chemical dissolution [48,49], then, by transfer printing the completed electronic systems from a rigid substrate to a flexible substrate.

In the first instance, to distinguish the normal tissue from abnormal, a highly flexible microscale piezoelectric device is developed [50]. Exploiting the alteration of the mechanical properties of tissue modulus in accordance with lesion expression [51–55], a novel minimally invasive probe offers quantitative agreement with clinical insights. The magnified schematic views of the specific structures are shown in Figure 2a. The thin, photolithographically patterned and defined triple-layer membrane, which is composed of Au/Cr, the piezoelectric material lead zirconate titanate (PZT), and Ti/Pt, is used as both an actuator and a sensor, simultaneously. Each of the components is transfer-printed onto the polyimide (PI) substrate from the donor wafer, regardless of the structure, in the form of either a free-standing device or integrated on a biopsy needle. After encapsulating the triple layer with PI, another layer of PI is used for encapsulation, after the deposition of the interconnecting Au lines. Figure 2b shows the images of the freestanding probe that has been placed into a biological tissue environment and the device has been wrapped around the surface of a conventional biopsy needle. The bottom right magnified image in Figure 2b designates the point of the actuator/sensor region on the needle. The fundamental device operation is based on the piezoelectric effect [56–60]. When the device achieves conformal contact with the tissue interface, the induced potential results in the piezoelectric strain at the actuator. The tissue that exhibits the deformation transfers this force to the sensor, and then, a portion of the strain force is changed into a certain value of voltage, thereby realizing the measurement of the tissue modulus. Figure 2c shows the results of the tissue modulus determination that has been conducted in human organs, such as fresh lung, adrenal gland, and fresh cirrhotic liver, as well as in a hepatocellular carcinoma. As estimated, Figure 2c provides the information that the cirrhotic liver maintains an average modulus of approximately 10 kPa, while the cancerous hepatocellular tissue reaches a peak modulus of approximately 23 kPa. The modulus graph in Figure 2d also shows the different modulus values that have been acquired between the fresh tissue and the abnormal tissue regions in the human thyroid and a formalin-treated kidney. In this example, the devices offer an underwork for the modulus-measuring platforms for biopsy guidance, based on magnetic resonance elastography [61–65].

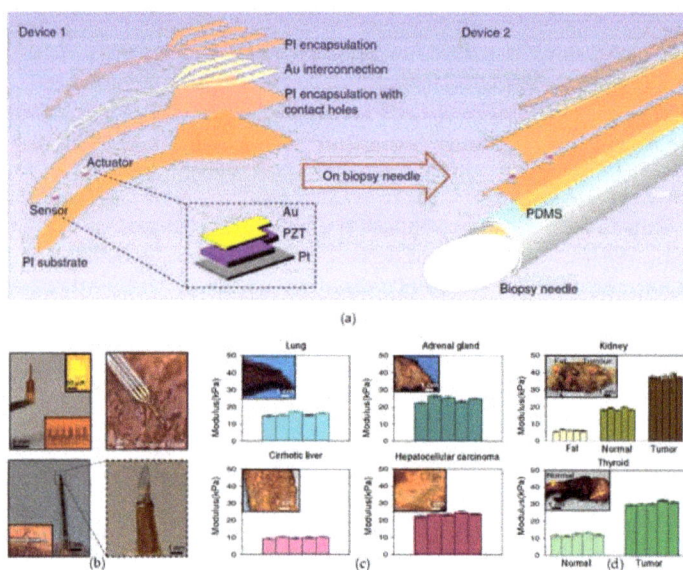

Figure 2. Flexible, piezoelectric tissue modulus recording probes: (**a**) Illustrations of two main piezoelectric material lead zirconate titanate (PZT)-based modulus probes, the free-standing device (left) and the conformally attached device on a biopsy needle (right); (**b**) pictures of self-standing devices (top) and the device on a biopsy needle (bottom), an inset picture (top, upper-right) designates a sensor/actuator pair, and the other inset picture (top, down-right) shows a row of completed designs. The magnified picture (bottom) shows specific actuator/sensor sites; (**c,d**) the measured modulus results for human tissues in a variety of organs using the PZT-based probe, the inset pictures of the upper-left side of the respective graphs show the optical images of organs in which modulus sensing is conducted. Reproduced with permission from 2018 Nature [50].

2.5. Novel Biomedical Electronics—Implantable, Soft Electronic Systems for Optical Stimulation

The top-down approach for nanopatterning can also be extensively applied to electrophysiology mapping [66–69] and electrical/optical stimulating systems [70–76]. Recently, opto-genetics have been extensively studied because of its delicate controllability of neural activity by selectively modifying channel-rhodopsin-treated genes with light. An entirely seamless implantable optoelectronic device, which lets axons react directly in accordance with light stimulation, is introduced in Figure 3a [77]. The radio-frequency (RF) harvesting unit rectifies the signals that have been received from the transmitter and routes the output current for the optical energy (LED). Then, an interconnected serpentine-structured Ti/Au antenna layer diminishes the resonant frequency while maintaining a wide-bandwidth requisite for efficient energy harvest at certain frequencies. The fabrication process for the device is initiated from the spin-casting of polyimide (PI) onto polymethyl methacrylate (PMMA) that has been coated on a glass surface. Then, a bilayer of Ti/Au is deposited by an e-beam evaporator and it is photolithographically patterned into the serpentine structure to form an antenna. After encapsulating the completed device with polydimethylsiloxane (PDMS), this component is immersed in an acetone solution so as to dissolve the PMMA layer. Finally, the thin free-standing composite optogenetic device is released from the glass, which results in a soft, flexible property. This flexible, stretchable state optimizes the conformal contact between the surface of the tissue and the device. An optogenetic control experiment is conducted, mainly in two parts, namely, underneath the gluteus maximus muscle and in the epidural space in the lumbar spinal cord of mice, as shown in Figure 3b,c. In order to determine the device functionality as an optogenetic platform, an experiment of Ch R2 activation for the nociceptive reaction of mice with spontaneous pain expression, following space abhorrence, is conducted. Figure 3d,e demonstrates the nociceptive pathways via the LED stimulation of the device that is fully implanted at the sciatic nerve and in the epidural space in the Ch R2 expressive mice, respectively. A rotarod test is conducted so as to identify the motor activity of the experimental rodent that has been affected by the performance of the LED stimulator (Figure 3i). The result proves that there is no change in the balance of the device. For specific experimental data, both mice expressing the Ch R2 and the control units are introduced to a y-shaped maze, where the mice can arbitrarily move to either side of the maze. In particular, at the left side of the passage in the y-shaped maze, there is an RF antenna that turns on the implanted LED, as illustrated in Figure 3h. Figure 3f,g show the number of nociceptive expressions and the trend of space abhorrence checking times of staying by the Advillin-Ch R2 for sciatic nerve-implanted mice, SNS-Ch R2 in the epidural space of spinal cord of mice, and the control units, according to the LED illumination. This result shows the expected correlation for stimulation, which presents the potential for the clinical device as a novel optogenetic therapy system [78,79] with great advantages, such as physical tether-free and external-feature-free characteristics.

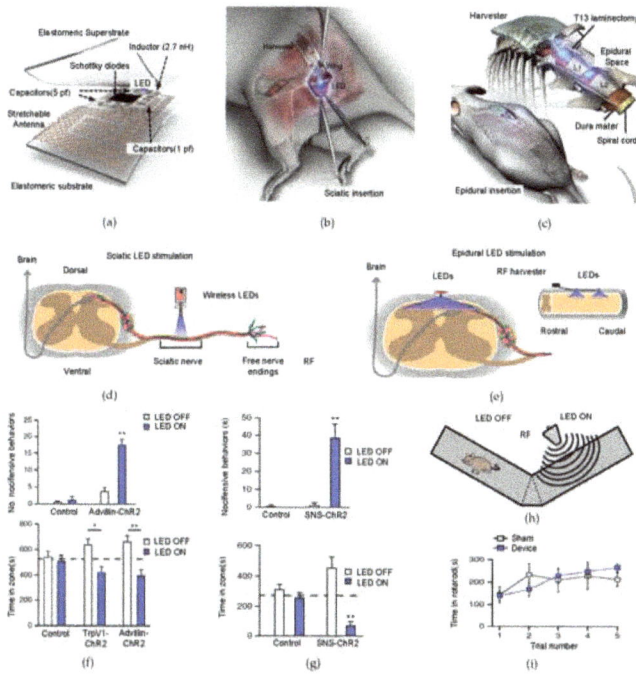

Figure 3. Fully implantable, flexible system for a wireless optogenetic stimulator with related experiments. (**a**) Schematic of a fully implantable, flexible electronic system for a wireless optogenetic stimulator; (**b**) images of the sciatic insertion surgery to implant the completed device into the spinal cord. The wing-shaped light emitting diode (LED) extension site passes from the gluteus maximus to the sciatic nerves; (**c**) a similar implantation is conducted at the epidural space in the spinal cord; (**d**,**e**) an illustration of the nociception path with an LED stimulator in a sciatic nerve and epidural space; (**f**,**g**) graphs of number of adverse behaviors/time in y-maze observed by the Advillin-ChR2-treated rodents in accordance with the LED stimulation in the sciatic nerve and epidural space; (**h**) an illustration of the experimental y-maze. Using radio-frequency (RF) signal, pathways are separated by LED on/off environments; (**i**) the graph of a rotarod performance result. This graph shows that motor activity does not change the entire performance of the device. Reproduced with permission from 2015 Nature [77].

2.6. Novel Biomedical Electronics—Cardiac Patches for Electrical Sensing, Stimulation, and Drug Delivery

The method of releasing an active device from a rigid substrate to obtain device mechanical flexibility is efficiently used in multifunctional cardiac patches (Figure 4a) [80]. These novel cardiac patches are composed of three main parts, namely, an electronic system for sensing and stimulating, electroactive polymer-deposited electrode sites for chemical factor diffusion to control cellular function, and three-dimensional (3D) dense nanofiber scaffolds in the cardiac cell environment. A detailed fabrication process is illustrated in Figure 4b. The device fabrication begins with depositing nickel as a relief layer onto a silicon substrate. Then, the spin-casting and curing of SU-8 photoresist on the surface of the Ni layer occurs, which is followed by the Cr/Au metal lift-off process. Subsequently, a titanium nitride (TiN) layer is deposited on the gold pads by sputtering, so as to increase their surface area. The passivated components of the design are released from the previous silicon substrate, and the remaining free-standing, flexible structure is obtained by etching the Ni relief layer using nitric acid. Figure 4c is an image of the flexible device, which consists of 32 electrodes with an SU-8 mesh structure.

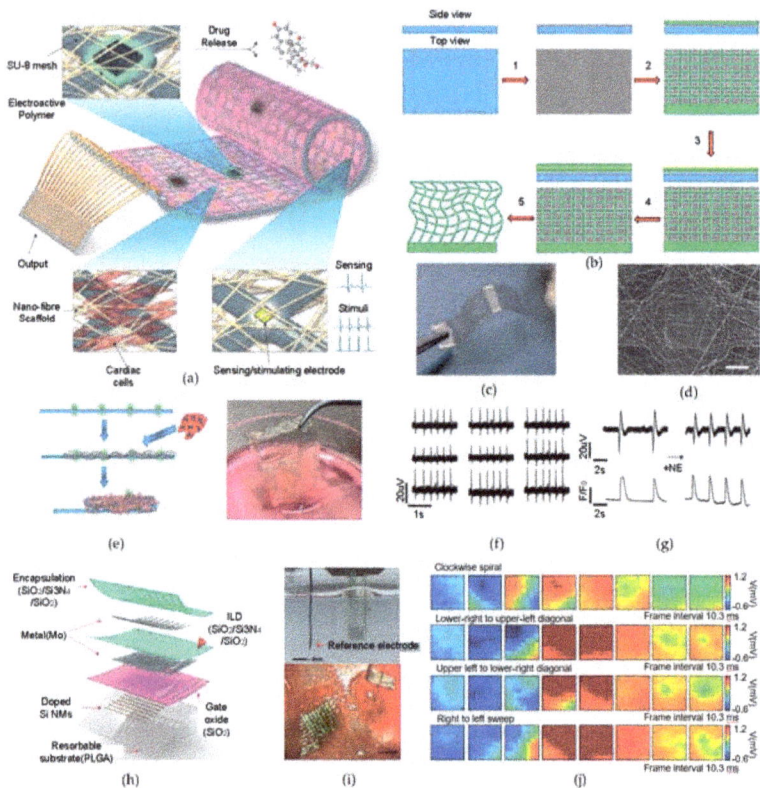

Figure 4. Multifunctional electronic system with biomaterial-based three-dimensional (3D) scaffold cardiac patches/multiplexed bio-resorbable silicon-based design for brain mapping. (**a**) Designed cardiac patches for recording electrical tissue activity/providing electrical stimulation/spatial releasing of biochemical factors; (**b**) the detailed fabrication process of the cardiac patches, the upper picture is the side view and the bottom picture is the top view of the device. The components include the temporary silicon wafer (blue), nickel layer (gray), SU-8 resist (green), and Au (yellow). (1) Deposition of a 20 nm nickel relief layer; (2) photolithography of SU-8 for the mesh structure; (3) patterning and Cr/Au deposition followed by the metal lift-off process and titanium nitride (TiN) layer deposition;

(4) coating the substrate with a uniform layer of SU-8, followed by photolithography, and curing and releasing the device from the substrate by etching the nickel layer with nitric acid; (**c**) a picture of a flexible device comprising 32 gold electrodes with a mesh of SU-8; (**d**) an SEM image of larger-sized electrodes in the cardiac patches with nanofiber-based 3D scaffolds; (**e**) a schematic of foldable patches with a conformal cell culture (left). The brown dots are electrodes that are used to release chemical factors, the gray serpentine lines are compounds of the 3D biomaterial scaffolds, and the pink lines are seeded cardiac cells. An image of the device in cell cultures (right); (**f**) the results of electrical signals from cardiac tissues recorded by nine gold electrodes; (**g**) images of increased signal frequency after treating with a neurotransmitter(norepinephrine) on the tissue (top) and integrated recordings through calcium imaging (bottom). Reproduced with permission from 2016 Nature [80]; (**h**) an enhanced schematic of the composition of an actively multiplexed brain-mapping sensor based on the highly doped SiNM; (**i**) a bio-resorbable device in phosphate buffer saline (PBS) with a reference electrode (top) and a magnified image of the implanted device at the left side of a rat hemisphere (bottom); (**j**) epilepsy signal-traced brain mapping results acquired by electrode channels from the device with various spikes. Reproduced with permission from 2016 Nature [81].

Fascinating cardiac cell cultures are optimized by the stitched-structure-like nanofiber-based biomaterial scaffolds. As shown in Figure 4d, the integration of the biomaterials gives the potential for on-demand drug release to control tissue functions without interfering with the seeding of cardiac cells. The schematic representation and picture of a device that has been cultivated in the cardiac site in Figure 4e indicates the process for the cell culture and the foldable scaffold structure with reliable cell viability. Electrical signal recordings that have been acquired by cardiomyocytes are done from nine gold electrodes in the device, shown in Figure 4f. As a result, all nine of the electrodes show similar results to the signals that have been generated by the cardiomyocytes. Figure 4g presents the results of the cardiac patch sensing for a signal frequency twice as high as that of administering norepinephrine and the integrated quantification recording, through calcium imaging.

2.7. Novel Biomedical Electronics—Bio-Resorbable, Ultraflexible Electronic Device for Transient Brain Mapping

In addition to the ability to sense temporary electrophysiological signals from the brain, a multiplexed bio-resorbable array of silicon transistor is introduced in Figure 4h [81]. This particular device is implanted over the cortical surface of a rat brain and functions for a programmed time period, resorbing in the body of the rat. Since the device only consists of biocompatible and biodegradable materials, a secondary surgery to extract the device is not required. After recording the brain activities for a predefined period of time, the device gradually dissolves in the body. The device is based on an actively multiplexed array with Si nano-membrane (SiNM) and N-metal-oxide-semiconductor (NMOS) transistors, which serve as the sites of the electrodes. The SiO_2 layer serves as the gate insulator, and the Mo serves as contacts for the source, drain, and gate. Two sets of triple layers of $SiO_2/Si_3N_4/SiO_2$ provide interlayer dielectric and encapsulation barriers. The fabrication begins with a silicon on insulator (SOI) wafer, which is doped with phosphorous to form the source and drain for the NMOS transistors. Then, the Si-NM layer is separated from the bulk Si wafer by dissolving a box SiO_2 layer, using hydrofluoric acid (HF). The Si-NM is then transfer-printed onto a thin PI/PMMA/Si substrate using a PDMS stamp, followed by the isolation of the SiNM, depositions of SiO_2 as the gate dielectric, and Mo for the source, drain, and gate contacts, to form the active matrix. The device is encapsulated and isolated with trilayer of $SiO_2/Si_3N_4/SiO_2$. Additionally, another layer of Mo is deposited via holes that are separating the columns and rows to form interconnects. Finally, the encapsulation of the device with another layer of a dilute PI and the formation of mesh structures by photolithography leads to transferring the entire active layer to a bio-resorbable substrate. After dissolving the PMMA layer using acetone to release the entire device, the mesh is consequently transferred onto the surface of the bio-resorbable poly-(lactic-co-glycolic acid) (PLGA) substrate to finalize the device fabrication. Soak testing is conducted in a phosphate buffer saline (PBS, pH 7.4) solution, as shown in Figure 4i (top). Figure 4i (bottom) shows an image of a conformally implanted bio-resorbable array in the left hemisphere and the control electrode in the right hemisphere of an anaesthetized rat brain, for in vivo recording. A series of eight movie frames from each epileptic spike activity, which are induced by applying picrotoxin, are shown in Figure 4j. The electrical mapping results of different forms of spikes (e.g., clockwise spiral, lower right to upper-left diagonal, upper left to lower-right diagonal, and right to left sweep) that have been recorded from the electrodes clearly demonstrate the spatio-temporally resolved neural propagation, thereby ensuring the potential for device use in the clinical arena or health-care systems.

3. Novel Devices Designed by Bottom-Up Nanofabrication

3.1. Introduction to Bottom-Up Approach

The components of flexible electronic devices at the nanoscale can be fabricated using the stacking of atoms or chemicals from the bottom to the top, referred to as bottom-up approaches. Such methods have distinct advantages in certain aspects compared to top-down approaches. The device fabrication process can be controlled precisely using bottom-up approaches and can be achieved at a relatively lower temperature, so that there is no risk of substrate deformation as a result of the applied heat [82]. There are various approaches to fabricate device components with novel materials using bottom-up approaches, such as growth methods [83–87], assembly methods [88–90], and chemical vapor deposition (CVD) [91–96]. Here, we introduce various applications for flexible electronic devices that have been fabricated by bottom-up approaches, such as a flexible transistor based on a MoSe$_2$ film, flexible light emitting diodes (LEDs) [93], a flexible touch screen [90], flexible strain sensors [83,88,89,97], and flexible temperature sensors [91].

3.2. Flexible Transistors Fabricated by Modified Chemical Vapor Deposition (mCVD)

As the most basic and necessary active device components in modern electronics, transistors have been developed with efforts focusing on reducing their size and power consumption. Moreover, researchers seek to find alternative materials for Si, which is mainly used as the material for conventional electronic devices, because the limitations in fabrication to achieve smaller physical dimensions give rise to low device packing densities. As a result of the mechanical rigidity of conventional transistors, there are limitations in applying them to flexible electronic devices. Therefore, flexible transistors have been widely studied. Here, transistors that are based on MoSe$_2$ films have been developed, which demonstrate high mobility and almost no difference in electrical performance, according to the I–V curves that have been obtained at different bending radii [92]. With the conventional chemical vapor deposition (CVD) process, large-scale MoS$_2$ films have poor electrical properties, such as field-effect mobilities lower than 15 cm$^2 \cdot$V$^{-1} \cdot$s^{-1} [98–101]. To form a MoSe$_2$ device layer with a high field-effect mobility, a modified CVD (mCVD) process that uses the polycrystalline compounds of MoSe$_2$ as the precursor to directly synthesize the product on SiO$_2$ is used. (Figure 5a) On the polyethylene terephthalate (PET) substrate, Ti/Al gate electrodes were deposited by electron beam evaporation and the dielectric material (SU-8 2000.5, MicroChem, Newton, MA, USA) was spin-coated. Then, a single crystalline multilayer of MoSe$_2$ flakes were transferred and the source-drain electrodes were defined to form flexible MoSe$_2$ transistors. High-mobility transistors based on a crystalline MoSe$_2$ film that have been grown on insulating SiO$_2$ can be fabricated with a field-effect mobility of 121 cm$^2 \cdot$V$^{-1} \cdot$s^{-1}. Figure 5b shows the fabricated flexible transistors with a bending radius of 'r'. Figure 5c shows the transfer curves of a MoSe$_2$ transistor in different bending regimes. The black, red, and green curves represent the MoSe$_2$ thin-film transistor (TFT) without bending and with bending, with radii of 10 mm and 5 mm, respectively. As a result, no significant changes or shifts in the transfer curves are found while changing the bending radius of the device. Therefore, the MoSe$_2$ TFTs that are fabricated by the modified CVD process are robust to induce mechanical stress, while maintaining their electrical performance.

❖ Bottom-up approaches

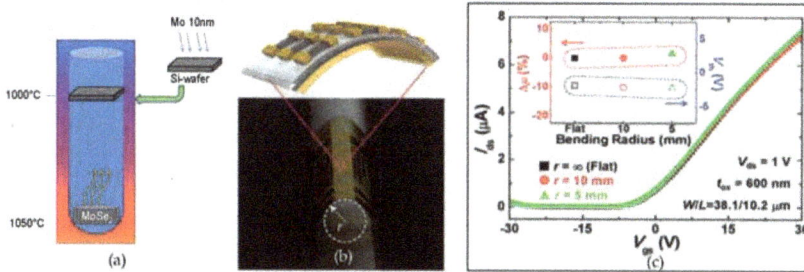

Figure 5. High-mobility transistors based on chemical vapor deposition (CVD)-grown MoSe₂; (**a**) schematic of synthesis of MoSe₂ film with modified CVD method; (**b**) optical image of bended MoSe₂ transistor; and (**c**) I–V characteristic of unbent and bent (up to 5 mm of radius) MoSe₂ transistor. Reproduced with permission from 2016 Advanced Materials [92].

3.3. Flexible Light Emitting Diodes (LEDs) and Flexible Touch Screen

Flexible light-emitting diodes (LEDs) open the path to conformal displays on complex curvilinear objects [102–104], healthcare devices [105,106], optoelectronic systems [107–109], instrumented surgical gloves [110], etc. Nanowires is a well-known material that is used to fabricate flexible LEDs. Among the various types of nanowires, nitride nanowires have remarkably good optoelectronics properties with excellent resistance to mechanical deformation [111]. Here, flexible LEDs based on nitride nanowires are introduced [93]. In particular, fully flexible blue LEDs using nanowires with a core/shell of InGaN/GaN that are grown via metalorganic chemical vapor deposition (MOCVD) have been demonstrated, as shown in Figure 6a. The fabricated blue LEDs (Figure 6b) show no degradation in light brightness, down to a bending radius of 3 mm and without encapsulation, while the conventional flexible LEDs require encapsulation barriers to protect the LEDs. In addition, two-layer bicolor flexible LEDs based on nanowires emitting blue and green light at the same time, have been demonstrated (Figure 6c). To fabricate the semitransparent green LED layer, which is the top layer of the two-layer bicolor flexible LED, on a thin metal shell of Ni/Au, GaN nanowire arrays are embedded in PDMS and are peeled off. Then, the whole layer is flipped to deposit Ti/Au, which serves as an arbitrary substrate. Since the AgNWs provide a high electrical conductivity, silver nanowires (AgNWs) are dispersed on the side opposite to the deposited Ti/Au (Figure 6d, 1–4). For the fabrication of the fully transparent blue LED, the fabrication process is similar to that of semitransparent LED. However, instead of the Ti/Au deposition, optically transparent AgNWs are dispersed on both sides of the PDMS-encapsulated layer (Figure 6d, 5–8). The fabricated two-layer flexible LED shows no significant performance degradation in I–V and electroluminescence (EL) characteristics at 3.5 mm and 2.5 mm bending radii, with repeated bending cycles.

The touch screen is another application of flexible electronic devices that have been fabricated via bottom-up approaches. Here, a solution-based self-assembled nanomesh, using aged gold nanowires (AuNWs), is fabricated [90]. As Figure 6e shows, keeping fresh AuNWs for 12 h before drop-casting forms aged AuNWs. The aged AuNWs self-assemble into bundles creating a continuous nanomesh structure with a pore size of 8–52 μm. While the steric hindrance of the fresh AuNWs is stronger than the wire-to-wire van der Waals force that maintains the ordered structure of nanomembrane, the steric hindrance and the van der Waals force balance is destroyed for the aged AuNWs, which creates bundles of AuNWs. As a result of the pores in the nanomesh from the aged AuNWs (Figure 6f), the nanomesh film is transparent, while the nanomembrane that is formed by fresh AuNWs is not. In addition, the nanomesh that is fabricated by the aged AuNWs is electrically conductive with a sheet resistance of 130.1 Ω^{-1}. The AuNW nanomesh can be transferred onto the polyethylene terephthalate (PET) with a sacrificial layer of a mask sheet. With the removal of the sacrificial mask by peeling, using tape or washing with ethanol, the patterned AuNW nanomesh is preserved on the PET (Figure 6g). The AuNW nanomesh can be utilized as an array of the pressure sensors for the flexible touch screen, as shown in Figure 6h. The nanomesh film shows a similar level of pressure sensitivity compared to that of commercial products [112].

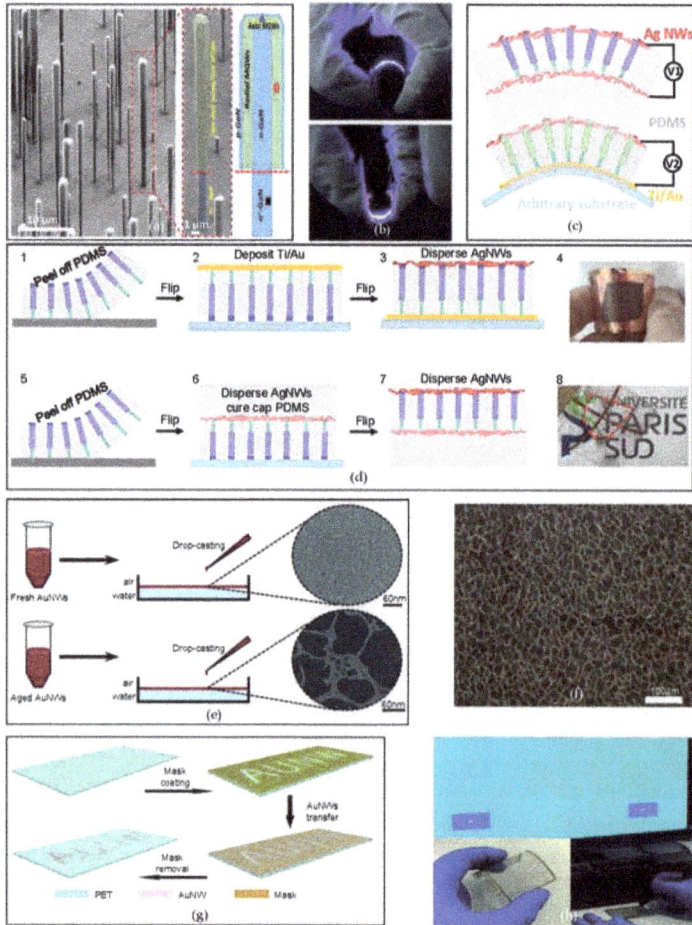

Figure 6. Flexible light emitting diodes based on vertically grown nitride nanowires/transparent and flexible nanomesh via self-assembly of ultrathin gold nanowires to fabricate a flexible touch panel. (**a**) An scanning electron microscope (SEM) image of a grown individual nanowire obtained with a tilt angle of 45°; (**b**) The flexibility of the fabricated nanowire blue LED; (**c**) A schematic of the bi-layer flexible LED with a blue LED for the top layer and a green LED for the bottom layer; (**d**) A schematic of the fabrication process for the semitransparent LED (1–4) and fully transparent LED (5–8), Reproduced with permission from 2015 Nano Letters [93]; (**e**) A transmission electron microscopy (TEM) image of a gold nanowires (AuNW) nanomembrane fabricated by a fresh AuNW solution and an SEM image of a AuNW mesh film fabricated from an aged AuNW solution; (**f**) An optical image of the AuNW mesh film with pores fabricated by the self-assembly method from an aged AuNW solution; (**g**) A schematic of the patterning of the AuNW mesh film on PET; and (**h**) A flexible touch panel fabricated using the AuNW mesh film. Reproduced with permission from 2016 Advanced Electronic Materials [90].

3.4. Novel Flexible Sensors—Strain Sensors

One of the well-known applications of the flexible electronic devices are the flexible sensors. Various types of substrate can be used for flexible strain sensors, from textiles to carbon-based materials. In order to enhance the electrical characteristics of a fabric-based flexible strain sensor, nanowires that are grown on the textile can be used [83]. Herein, a wireless flexible strain sensor is applied on a commercial textile via the assembly of hybrid carbon materials at the nanoscale and piezo-resistive ZnO nanowires (NWs) that have been grown on the commercial textile. The assembled hybrid carbon nanomaterials provide excellent properties in terms of the robust electrical performances against mechanical deformation from bending. On the pristine PET textile, carbon nanotubes (CNTs) and reduced graphene oxide (rGO) are coated so as to provide an excellent growth condition for the ZnO nanowires [113,114]. The coated hybrid carbon nanomaterials provide the continuous conductivity and durability against the externally applied mechanical strain. Then, the ZnO-based texture-type strain sensor is encapsulated with a thin PDMS layer (Figure 7a). Figure 7b,c show the current density change with the bending and the release of the fabricated flexible textile strain sensor. The plots in Figure 7c shows the flexible textile sensor monitoring the repeated bending and release of an arm up to bending angles of 60° and 120°, respectively. The assembled ZnO-3 on the PET substrate textile, which is hybridized with carbon nanomaterials, has a twice-higher gauge factor (GF) of 7.64, than that of ZnO-3 that has been assembled on a plain conventional PET film. This result implies that using the fabric-like structure as the substrate for strain sensors culminates in high-pressure sensitivity and stability at the same time.

Skin-like electronics, known as e-skins, have been widely studied because of their wide applications, such as health monitoring, medical implantation, and integration of sensors. A wide range of novel materials can be used to fabricate e-skins. Among those materials, graphene is currently one of the most widely used materials to fabricate the active devices [115–118]. Here, a highly sensitive flexible strain sensor for e-skins is developed using an ultrathin sensitive graphene film that is formed through the self-assembly method [89]. The self-assembly method uses the specific affinity between two molecules [119–121]. The molecules that are used to form a self-assembled layer have strong interaction forces, which are noncovalent interactions, so that a monolayer can be formed tightly. Van der Waals [122,123], hydrogen bonding [124], and π-π interactions [125,126] are the representative examples of noncovalent interactions that are used in the self-assembly method. The Marangoni effect [127–129], which is the phenomenon of transferring a group because of the difference of a gradient between two different fluids' surface tensions, is used to form an ultrathin graphene film (UGF). Figure 7d shows the schematic of how the ultrathin graphene film (UGF) is formed by the self-assembly method. Graphene flakes, with a thickness of 2.5 nm, are injected on the surface of the deionized (DI) water, and ethanol is spread on the surface of the DI water. As a result of the Marangoni effect, the ethanol with graphene flakes tend to move toward the high surface tension area, where the DI water is richer than ethanol. Then, the transferred graphene flakes collide and compact together via π-π interactions, forming a large area of UGF, as shown in Figure 7e. The whole process of forming a large-area UGF of 150 cm^2 takes only 5 s. The advantage of using the self-assembly process that is introduced here, is that the self-assembly process can form the uniform UGF, while the other methods, such as drop-casting and spin-coating, cannot. With the UGF that is formed by the self-assembly method, the strain sensor can be fabricated. As shown in Figure 7f, the UGF is transferred onto the mask, which is attached on the PDMS slab. After removing the mask to form 8 × 8 pixels, the electrodes are transferred to create an electronic skin with the tactile sensors. The UGF that is formed by the self-assembly process shows an extremely high gauge factor (GF) of 1037 at a low strain (2%). The resistance of the individual graphene flakes hardly changes because of the stable crystal structure. However, using the tunneling effect [130] of the overlapped intersection area of the self-assembled graphene flakes, the exceptionally high GF that has been shown in the result, can be achieved. Figure 7g (top) shows the performance of the pressure sensor array, wrapped around the arm. As the strain is applied from the fingers, the spatially resolved pressure sensing can be monitored, as shown in Figure 7g (bottom).

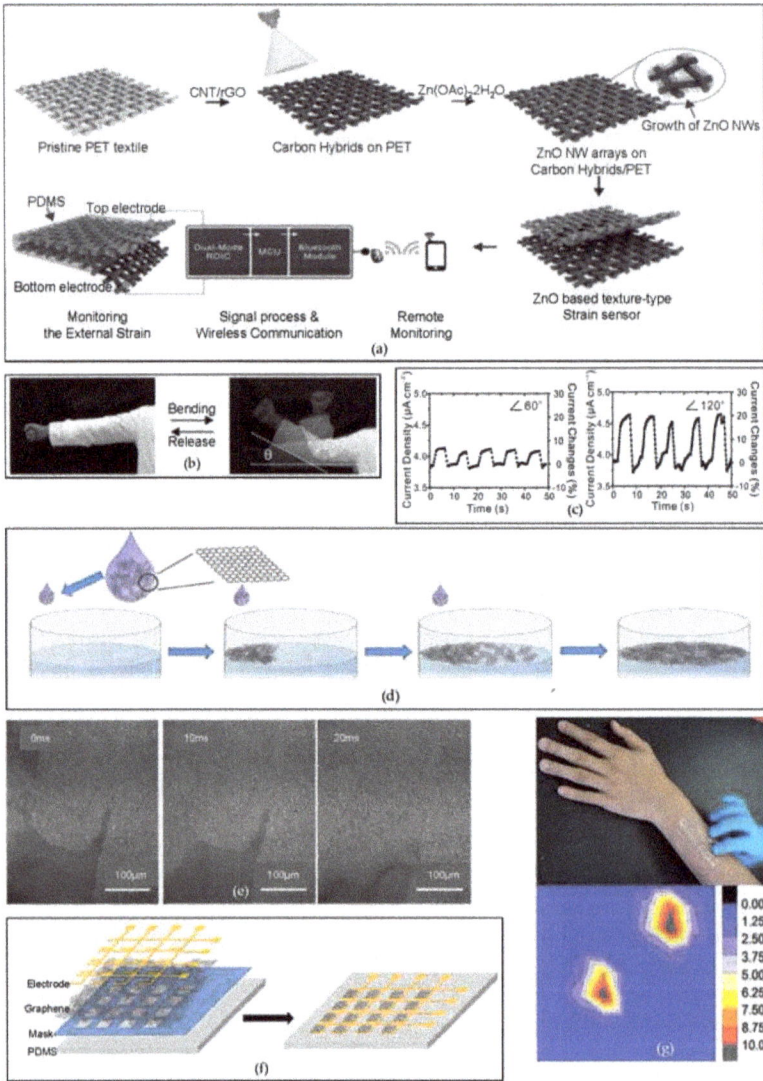

Figure 7. Flexible textile sensor based on a hybrid of carbon materials as the substrate/Strain sensors based on ultrathin graphene films. (**a**) A schematic of the fabrication of a flexible textile strain sensor generated via the growth of ZnO nanowires; (**b,c**) The bending and releasing of the cloth made by a flexible textile strain sensor and the corresponding results according to the different bending angles of the arm; Reproduced with permission from 2016 Advanced Functional Materials [83]; (**d**) Graphene flakes forming the ultrathin graphene film (UGF) via the π-π interactions among each other; (**e**) Optical images of the formation of UGF taken in 20 ms by a high-speed charge-coupled device (CCD) camera; (**f**) A schematic of a strain sensor fabricated from the self-assembled UGF; (**g**) A strain sensor applied on the forearm and the real-time response map of pressure applied on the strain sensor. Reproduced with permission from 2016 Advanced Materials [89].

Graphene shows a remarkable electrical conductivity, mechanical strength, flexibility, and optical transparency. With these properties, using the graphene and microstructured graphene that has been obtained by the layer via a layer-by-layer (LBL) assembly method [131–134], an ultrasensitive pressure sensor can be developed. The advantage of the LBL assembly method is the thickness of the LBL assembled layer can be easily controlled through the adjustment of the number of layers and the shape of the LBL assembled layer via uniformly distributed molecules on each layer. As a result of the geometrical and self-limiting properties, the surface structure of the ultrathin film that has been formed by the LBL assembly method can be designed in a variety of shapes, while this is not possible with other methods, such as spin-coating or drop-casting [88]. Figure 8a shows the fabrication process of the ultrasensitive flexible tactile sensor. With a KOH-etched Si master mold, pyramid shapes of PDMS can be fabricated. Graphene oxide (GO) from graphite oxide is well suited for the LBL assembly because the GO sheet has abundant negative charges with the hydroxy and carboxyl groups. After depositing the GO sheets on the microstructured PDMS, which improves the sensitivity via structure, the GO layers are treated with hydrazine vapor to form reduced graphene oxide (rGO), which consists of electrically conductive graphene sheets [135,136]. The rGO sheets are sandwiched between the PDMS and an indium tin oxide (ITO)-coated PET film to construct the pressure sensor unit. The tips of the pyramid shapes of the rGO contact the ITO-coated PET film, which increase the sensitivity of the device. Figure 8b shows the relative resistance changes ($\Delta R/R0$) of the pressure changes. The microstructured film is highly sensitive over the unstructured film. The pressure-response curves of the microstructured film can be subdivided into two parts, namely, highly sensitive at lower pressures with a range of 0–100 kPa, and a saturated sensitivity segment with a range above 100 kPa. At the lower pressure range (0–100 kPa), the pressure sensitivity slope is -5.53 kPa^{-1}, and at the saturation in the sensitivity range (>100 kPa), the pressure sensitivity slope is -0.01 kPa^{-1}. This means the highly sensitive flexible tactile sensor is ultrasensitive at the lower pressure range of 0–100 kPa, with an ultrafast response time of 0.2 ms. As Figure 8c shows, the ultrasensitive at a lower pressure range (0–100 kPa) pressure sensor for e-skin has been developed.

In another example of a strain sensor for e-skin, the arrays of the stretchable transistors are fabricated with components of polystyrene-block-poly(ethylene-ran-butylene)-block-polystyrene (SEBS) as a flexible dielectric, a conjugated polymer/elastomer phase separation induced elasticity (CONPHINE) film as the semiconductor, and spray-coated carbon nanotubes (CNTs) for electrodes [97]. Figure 8d shows the fabrication processes of the arrays of the stretchable transistors. Rigid Si/SiO$_2$ is used as the bottom layer to fabricate the arrays, and on the top, water soluble dextran, which is used as a sacrificial layer to separate the active layer from the rigid Si/SiO$_2$ layer, is deposited via spin-coating. Then, a stretchable dielectric, stretchable semiconductor, and stretchable conductor (CNTs) are patterned. The SEBS is laminated on to function as the flexible substrate, and the CNTs are patterned to form the gate electrodes. Figure 8e shows the schematic of the formed stretchable transistor. The completed arrays of the stretchable transistors show a slight difference in the carrier mobility when applying external stresses vertically and horizontally (Figure 8f). The average charge-carrier mobility from the array of transistors that were recorded is 0.821 ± 0.105 cm$^2 \cdot$V$^{-1} \cdot$s^{-1}, with the highest value of 1.11 cm$^2 \cdot$V$^{-1} \cdot$s^{-1}, an on/off current ratio of 10^4, and a low operation voltage of 10 V. In addition, this stretchable transistor array consumes little power, which suggests its potential for use as a self-powered e-skin (Figure 8f,g). Figure 8h shows the stretchable array of the transistors attached onto a palm. The device can detect very small pressure changes with a high spatially resolved pattern and sensitivity, even when an artificial lady bug sits on the device.

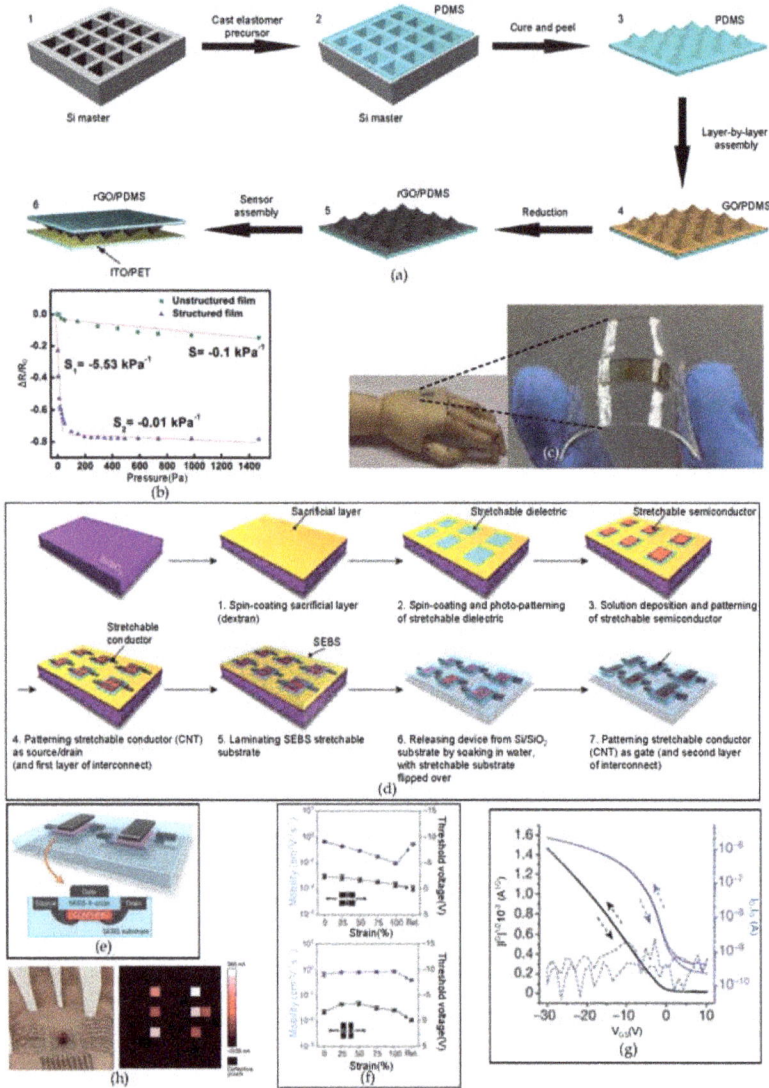

Figure 8. Microstructured graphene arrays for sensitive flexible tactile sensors/electronic skin based on a stretchable transistor array. (**a**) A schematic of the fabrication of a flexible tactile sensor device via the layer-by-layer (LBL) assembly method. An ultrathin graphene film that detects applied pressure is deposited on the PDMS film; (**b**) the pressure response curves of unstructured and structured rGO/PDMS films. The slope of the relative resistance change is -5.53 kPa^{-1} at 100 Pa, which shows very sensitive changes at lower pressures; (**c**) an image of the tactile senor applied on the flexible substrate. Reproduced with permission from 2014 Small [88]. (**d**) a schematic of the fabrication process for arrays of flexible transistors; (**e**) a magnified view of a flexible transistor exhibiting high performance; (**f**) the mobility and threshold voltage changes when applying strain from 0% to 100%; (**g**) the I–V characteristics of fabricated transistors, which have almost no hysteresis; (**h**) stretchable arrays of transistors attached to the human palm with a small artificial lady bug with six legs and the corresponding result from the stretchable sensor. Reproduced with permission from 2018 Nature [97].

3.5. Novel Flexible Sensors—Temperature Sensor

Monitoring the body temperature is important since numerous diseases can be characterized by abnormal changes of body temperature. Therefore, a precise and continuous temperature monitoring device is required. Recent work presents a stretchable and flexible temperature senor in the form of electronic skin, with a stable temperature sensing performance when an external strain is applied [91]. The device consists of a 5 × 5 single-walled carbon nanotube (SWCNT) TFT layer, two liquid metal interconnection layers, and a layer of a 5 × 5 array of temperature sensors. On layer 1, the array of the SWCNT TFTs are on a PET film, having been fabricated via a CVD process and transferred. Layer 2 and layer 3 are microchannel Ecoflex layers, which are formed on a microstructured mold. Layer 4 is the temperature sensing layer with polyaniline temperature sensors on a PET film. Four different layers are electrically connected via a liquid metal that is a compound form of gallium, indium, and tin (Figure 9a,b). Since the device consists of the thin soft elastomer (Ecoflex) and ultrathin active components, the entire device can be stretched up to a strain of 30%, without device fracturing or performance degradation. The completely fabricated temperature sensor can sense both heating and cooling through the normalized resistance change, with respect to the temperature change. There is no hysteresis found in the normalized resistance change for the heating and cooling. Figure 9c shows that the temperature sensor array can be bent up to a radius of 14 mm, while maintaining its electrical performances.

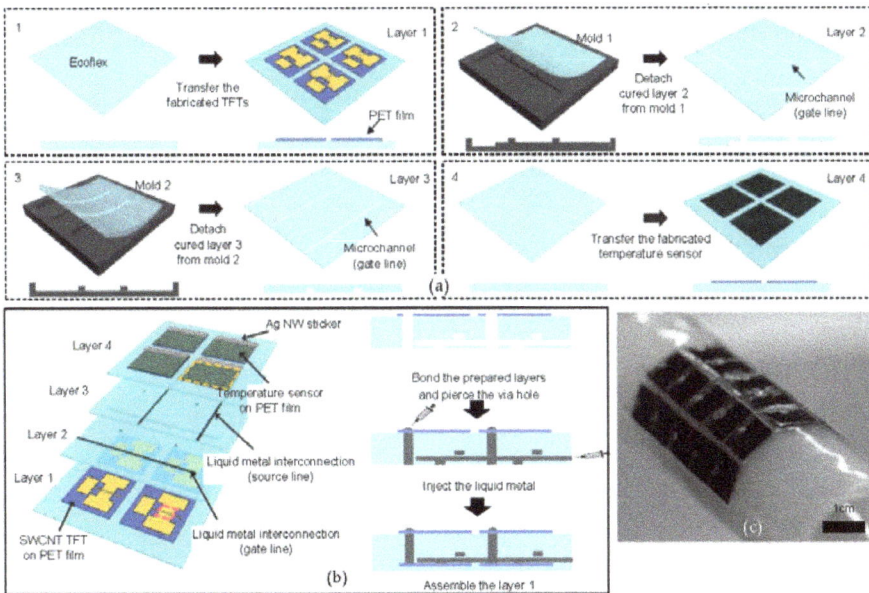

Figure 9. Stretchable temperature sensor array for electronic skin. (**a**) A schematic of the fabrication of the flexible and stretchable active matrix temperature sensor array; (**b**) the vertical schematic view of the sensor (left) and injected liquid metal for gate/source lines; (**c**) an optical image of the bent sensor with a radius of 14 mm. Reproduced with permission from 2016 Advanced Materials [91].

4. Conclusions

This review summarizes the most recent developments in the novel fabrication techniques of flexible and stretchable electronics. Modulus-sensing biopsy probes, implantable optoelectronic systems, multifunctional cardiac patches, and bio-resorbable transient brain-mapping electronic designs are introduced in the contents of the top-down approach. Then, Au nanomembrane by self-assembly method for touch screen, flexible textile strain sensing system, graphene-based tactile sensors, stretchable transistor arrays, and temperature sensor designs are also demonstrated in the bottom-up contents. The graphene transistor, photonic devices, $MoSe_2$ transistor, and LEDs are also discussed in the introductory contents. Detailed fabrication methods that have been previously mentioned, such as transfer printing with replica molding, transfer and releasing processes, growth, and self-assembly shows that most of the processes are available commercially at low-cost with high quality. These approaches are important because they strongly suggest solutions for high-temperature processes, which are susceptible for the flexible substrate and high scalability in a short time that conventional fabrication processes cannot achieve. Various techniques that have been introduced in this review show the greatest potentials of flexible electronics used as clinical issues, human healthcare system, and multifunctional sensing devices with optimistic practicality and reliability.

Author Contributions: K.K., Y.U.C., and K.J.Y. collected the data, contributed to the scientific discussions, and co-wrote the manuscript.

Acknowledgments: K.K., Y.U.C., and K.J.Y. acknowledge the support from the National Research Foundation of Korea (Grant No. NRF-2017M1A2A2048904) and the Yonsei University Future-Leading Research Initiative of 2017 (RMS22017-22-00).

Conflicts of Interest: The authors declare no competing financial interests.

References

1. Gates, B.D.; Xu, Q.; Stewart, M.; Ryan, D.; Willson, C.G.; Whitesides, G.M. New approaches to nanofabrication: Molding, printing, and other techniques. *Chem. Rev.* **2005**, *105*, 1171–1196. [CrossRef] [PubMed]
2. Chen, Y.; Pepin, A. Nanofabrication: Conventional and nonconventional methods. *Electrophoresis* **2001**, *22*, 187–207. [CrossRef]
3. Gates, B.D.; Xu, Q.; Love, J.C.; Wolfe, D.B.; Whitesides, G.M. Unconventional nanofabrication. *Annu. Rev. Mater. Res.* **2004**, *34*, 339–372. [CrossRef]
4. Quake, S.R.; Scherer, A. From micro-to nanofabrication with soft materials. *Science* **2000**, *290*, 1536–1540. [CrossRef] [PubMed]
5. Yang, L.; Akhatov, I.; Mahinfalah, M.; Jang, B.Z. Nano-fabrication: A review. *J. Chin. Inst. Eng.* **2007**, *30*, 441–446. [CrossRef]
6. Wang, X.; Lu, X.; Liu, B.; Chen, D.; Tong, Y.; Shen, G. Flexible energy-storage devices: Design consideration and recent progress. *Adv. Mater.* **2014**, *26*, 4763–4782. [CrossRef] [PubMed]
7. Chen, H.; Cong, T.N.; Yang, W.; Tan, C.; Li, Y.; Ding, Y. Progress in electrical energy storage system: A critical review. *Prog. Nat. Sci.* **2009**, *19*, 291–312. [CrossRef]
8. Dong, L.; Xu, C.; Li, Y.; Huang, Z.-H.; Kang, F.; Yang, Q.-H.; Zhao, X. Flexible electrodes and supercapacitors for wearable energy storage: A review by category. *J. Mater. Chem. A* **2016**, *4*, 4659–4685. [CrossRef]
9. Stoppa, M.; Chiolerio, A. Wearable electronics and smart textiles: A critical review. *Sensors* **2014**, *14*, 11957–11992. [CrossRef] [PubMed]
10. Zeng, W.; Shu, L.; Li, Q.; Chen, S.; Wang, F.; Tao, X.M. Fiber-based wearable electronics: A review of materials, fabrication, devices, and applications. *Adv. Mater.* **2014**, *26*, 5310–5336. [CrossRef] [PubMed]
11. Tao, X. *Wearable Electronics and Photonics*; Elsevier: New York, NY, USA, 2005.
12. Mukhopadhyay, S.C. Wearable sensors for human activity monitoring: A review. *IEEE Sens. J.* **2015**, *15*, 1321–1330. [CrossRef]
13. Chen, Y.; Au, J.; Kazlas, P.; Ritenour, A.; Gates, H.; McCreary, M. Electronic paper: Flexible active-matrix electronic ink display. *Nature* **2003**, *423*, 136. [CrossRef] [PubMed]
14. Chang, P.-L.; Wu, C.-C.; Leu, H.-J. Investigation of technological trends in flexible display fabrication through patent analysis. *Displays* **2012**, *33*, 68–73. [CrossRef]
15. Chung, I.-J.; Kang, I. Flexible display technology—Opportunity and challenges to new business application. *Mol. Cryst. Liq. Cryst.* **2009**, *507*, 1–17. [CrossRef]
16. Biswas, A.; Bayer, I.S.; Biris, A.S.; Wang, T.; Dervishi, E.; Faupel, F. Advances in top–down and bottom–up surface nanofabrication: Techniques, applications & future prospects. *Adv. Colloid Interface Sci.* **2012**, *170*, 2–27. [PubMed]

17. Khanna, V.K. Top-down nanofabrication. In *Integrated Nanoelectronics*; Springer: Berlin, Germany, 2016; pp. 381–396.

18. Mijatovic, D.; Eijkel, J.C.; van den Berg, A. Technologies for nanofluidic systems: Top-down vs. Bottom-up—A review. *Lab Chip* **2005**, *5*, 492–500. [CrossRef] [PubMed]

19. Molnár, G.; Cobo, S.; Real, J.A.; Carcenac, F.; Daran, E.; Vieu, C.; Bousseksou, A. A combined top-down/bottom-up approach for the nanoscale patterning of spin-crossover coordination polymers. *Adv. Mater.* **2007**, *19*, 2163–2167. [CrossRef]

20. Cavallini, M.; Facchini, M.; Massi, M.; Biscarini, F. Bottom–up nanofabrication of materials for organic electronics. *Synth. Met.* **2004**, *146*, 283–286. [CrossRef]

21. Khanna, V.K. Bottom-up nanofabrication. In *Integrated Nanoelectronics*; Springer: Berlin, Germany, 2016; pp. 397–417.

22. Lu, W.; Lieber, C.M. Nanoelectronics from the bottom up. *Nat. Mater.* **2007**, *6*, 841–850. [CrossRef] [PubMed]

23. Levinson, H.J. *Lithography Process Control*; SPIE Optical Engineering Press: Bellingham, WA, USA, 1999.

24. Watt, F.; Bettiol, A.; Van Kan, J.; Teo, E.; Breese, M. Ion beam lithography and nanofabrication: A review. *Int. J. Nanosci.* **2005**, *4*, 269–286. [CrossRef]

25. Bondur, J.A. Dry process technology (reactive ion etching). *J. Vac. Sci. Technol.* **1976**, *13*, 1023–1029. [CrossRef]

26. Jansen, H.; Gardeniers, H.; de Boer, M.; Elwenspoek, M.; Fluitman, J. A survey on the reactive ion etching of silicon in microtechnology. *J. Micromech. Microeng.* **1996**, *6*, 14. [CrossRef]

27. Yih, P.; Saxena, V.; Steckl, A. A review of SiC reactive ion etching in fluorinated plasmas. *Phys. Status Solidi (b)* **1997**, *202*, 605–642. [CrossRef]

28. Zardetto, V.; Brown, T.M.; Reale, A.; Di Carlo, A. Substrates for flexible electronics: A practical investigation on the electrical, film flexibility, optical, temperature, and solvent resistance properties. *J. Polym. Sci. B Polym. Phys.* **2011**, *49*, 638–648. [CrossRef]

29. Carlson, A.; Bowen, A.M.; Huang, Y.; Nuzzo, R.G.; Rogers, J.A. Transfer printing techniques for materials assembly and micro/nanodevice fabrication. *Adv. Mater.* **2012**, *24*, 5284–5318. [CrossRef] [PubMed]

30. Hines, D.; Ballarotto, V.; Williams, E.; Shao, Y.; Solin, S. Transfer printing methods for the fabrication of flexible organic electronics. *J. Appl. Phys.* **2007**, *101*, 024503. [CrossRef]

31. Lee, C.H.; Kim, D.R.; Zheng, X. Fabricating nanowire devices on diverse substrates by simple transfer-printing methods. *Proc. Natl. Acad. Sci. USA* **2010**, *107*, 9950–9955. [CrossRef] [PubMed]

32. Lee, C.H.; Kim, D.R.; Zheng, X. Transfer printing methods for flexible thin film solar cells: Basic concepts and working principles. *ACS Nano* **2014**, *8*, 8746–8756. [CrossRef] [PubMed]

33. Sun, Y.; Rogers, J.A. Fabricating semiconductor nano/microwires and transfer printing ordered arrays of them onto plastic substrates. *Nano Lett.* **2004**, *4*, 1953–1959. [CrossRef]

34. Ji, S.; Liu, C.-C.; Liu, G.; Nealey, P.F. Molecular transfer printing using block copolymers. *ACS Nano* **2009**, *4*, 599–609. [CrossRef] [PubMed]

35. Loo, Y.-L.; Willett, R.L.; Baldwin, K.W.; Rogers, J.A. Interfacial chemistries for nanoscale transfer printing. *J. Am. Chem. Soc.* **2002**, *124*, 7654–7655. [CrossRef] [PubMed]

36. Meitl, M.A.; Zhu, Z.-T.; Kumar, V.; Lee, K.J.; Feng, X.; Huang, Y.Y.; Adesida, I.; Nuzzo, R.G.; Rogers, J.A. Transfer printing by kinetic control of adhesion to an elastomeric stamp. *Nat. Mater.* **2006**, *5*, 33. [CrossRef]

37. Song, D.; Mahajan, A.; Secor, E.B.; Hersam, M.C.; Francis, L.F.; Frisbie, C.D. High-resolution transfer printing of graphene lines for fully printed, flexible electronics. *ACS Nano* **2017**, *11*, 7431–7439. [CrossRef] [PubMed]

38. Espinha, A.; Dore, C.; Matricardi, C.; Alonso, M.I.; Goñi, A.R.; Mihi, A. Hydroxypropyl cellulose photonic architectures by soft nanoimprinting lithography. *Nat. Photonics* **2018**, *1*. [CrossRef]

39. Ishikawa, A.; Shffiata, T. Cellulosic chiral stationary phase under reversed-phase condition. *J. Liq. Chromatogr. Relat. Technol.* **1993**, *16*, 859–878. [CrossRef]

40. O'brien, T.; Crocker, L.; Thompson, R.; Thompson, K.; Toma, P.; Conlon, D.; Feibush, B.; Moeder, C.; Bicker, G.; Grinberg, N. Mechanistic aspects of chiral discrimination on modified cellulose. *Anal. Chem.* **1997**, *69*, 1999–2007. [CrossRef] [PubMed]

41. Revol, J.-F.; Godbout, L.; Dong, X.-M.; Gray, D.G.; Chanzy, H.; Maret, G. Chiral nematic suspensions of cellulose crystallites; phase separation and magnetic field orientation. *Liq. Cryst.* **1994**, *16*, 127–134. [CrossRef]

42. Shopsowitz, K.E.; Hamad, W.Y.; MacLachlan, M.J. Chiral nematic mesoporous carbon derived from nanocrystalline cellulose. *Angew. Chem. Int. Ed.* **2011**, *50*, 10991–10995. [CrossRef] [PubMed]

43. Qin, D.; Xia, Y.; Whitesides, G.M. Soft lithography for micro-and nanoscale patterning. *Nat. Protoc.* **2010**, *5*, 491–502. [CrossRef] [PubMed]

44. Rogers, J.A.; Nuzzo, R.G. Recent progress in soft lithography. *Mater. Today* **2005**, *8*, 50–56. [CrossRef]

45. Rolland, J.P.; Hagberg, E.C.; Denison, G.M.; Carter, K.R.; De Simone, J.M. High-resolution soft lithography: Enabling materials for nanotechnologies. *Angew. Chem. Int. Ed.* **2004**, *43*, 5796–5799. [CrossRef] [PubMed]

46. Whitesides, G.M.; Ostuni, E.; Takayama, S.; Jiang, X.; Ingber, D.E. Soft lithography in biology and biochemistry. *Annu. Rev. Biomed. Eng.* **2001**, *3*, 335–373. [CrossRef] [PubMed]

47. Xia, Y.; Whitesides, G.M. Soft lithography. *Annu. Rev. Mater. Sci.* **1998**, *28*, 153–184. [CrossRef]

48. Demeester, P.; Pollentier, I.; De Dobbelaere, P.; Brys, C.; Van Daele, P. Epitaxial lift-off and its applications. *Semicond. Sci. Technol.* **1993**, *8*, 1124. [CrossRef]

49. Liao, W.-S.; Cheunkar, S.; Cao, H.H.; Bednar, H.R.; Weiss, P.S.; Andrews, A.M. Subtractive patterning via chemical lift-off lithography. *Science* **2012**, *337*, 1517–1521. [CrossRef] [PubMed]

50. Yu, X.; Wang, H.; Ning, X.; Sun, R.; Albadawi, H.; Salomao, M.; Silva, A.C.; Yu, Y.; Tian, L.; Koh, A. Needle-shaped ultrathin piezoelectric microsystem for guided tissue targeting via mechanical sensing. *Nat. Biomed. Eng.* **2018**, *2*, 165–172. [CrossRef]

51. Abramowitch, S.D.; Feola, A.; Jallah, Z.; Moalli, P.A. Tissue mechanics, animal models, and pelvic organ prolapse: A review. *Eur. J. Obstet. Gynecol. Reprod. Biol.* **2009**, *144*, S146–S158. [CrossRef] [PubMed]

52. Cowin, S.C.; Doty, S.B. *Tissue Mechanics*; Springer Science & Business Media: Berlin, Germany, 2007.

53. Discher, D.E.; Janmey, P.; Wang, Y.-L. Tissue cells feel and respond to the stiffness of their substrate. *Science* **2005**, *310*, 1139–1143. [CrossRef] [PubMed]

54. Hoyt, K.; Castaneda, B.; Zhang, M.; Nigwekar, P.; di Sant'Agnese, P.A.; Joseph, J.V.; Strang, J.; Rubens, D.J.; Parker, K.J. Tissue elasticity properties as biomarkers for prostate cancer. *Cancer Biomark.* **2008**, *4*, 213–225. [CrossRef] [PubMed]

55. Kenedi, R.; Gibson, T.; Evans, J.; Barbenel, J. Tissue mechanics. *Phys. Med. Biol.* **1975**, *20*, 699. [CrossRef] [PubMed]

56. Bassett, C.A.L. Biologic significance of piezoelectricity. *Calcif. Tissue Res.* **1967**, *1*, 252–272. [CrossRef]

57. Marino, A.A.; Becker, R.O.; Soderholm, S.C. Origin of the piezoelectric effect in bone. *Calcif. Tissue Res.* **1971**, *8*, 177–180. [CrossRef] [PubMed]

58. Ribeiro, C.; Sencadas, V.; Correia, D.M.; Lanceros-Méndez, S. Piezoelectric polymers as biomaterials for tissue engineering applications. *Colloids Surf. B Biointerfaces* **2015**, *136*, 46–55. [CrossRef] [PubMed]

59. Shamos, M.H.; Lavine, L.S. Piezoelectricity as a fundamental property of biological tissues. *Nature* **1967**, *213*, 267–269. [CrossRef] [PubMed]

60. Telega, J.J.; Wojnar, R. Piezoelectric effects in biological tissues. *J. Theor. Appl. Mech.* **2002**, *40*, 723–759.

61. Arda, K.; Ciledag, N.; Aktas, E.; Arıbas, B.K.; Köse, K. Quantitative assessment of normal soft-tissue elasticity using shear-wave ultrasound elastography. *Am. J. Roentgenol.* **2011**, *197*, 532–536. [CrossRef] [PubMed]

62. Kruse, S.; Smith, J.; Lawrence, A.; Dresner, M.; Manduca, A.; Greenleaf, J.F.; Ehman, R.L. Tissue characterization using magnetic resonance elastography: Preliminary results. *Phys. Med. Biol.* **2000**, *45*, 1579. [CrossRef] [PubMed]

63. Manduca, A.; Oliphant, T.E.; Dresner, M.; Mahowald, J.; Kruse, S.A.; Amromin, E.; Felmlee, J.P.; Greenleaf, J.F.; Ehman, R.L. Magnetic resonance elastography: Non-invasive mapping of tissue elasticity. *Med. Image Anal.* **2001**, *5*, 237–254. [CrossRef]

64. Ophir, J.; Cespedes, I.; Garra, B.; Ponnekanti, H.; Huang, Y.; Maklad, N. Elastography: Ultrasonic imaging of tissue strain and elastic modulus in vivo. *Eur. J. Ultrasound* **1996**, *3*, 49–70. [CrossRef]

65. Ophir, J.; Cespedes, I.; Ponnekanti, H.; Yazdi, Y.; Li, X. Elastography: A quantitative method for imaging the elasticity of biological tissues. *Ultrason. Imaging* **1991**, *13*, 111–134. [CrossRef] [PubMed]

66. Gross, R.E.; Krack, P.; Rodriguez-Oroz, M.C.; Rezai, A.R.; Benabid, A.L. Electrophysiological mapping for the implantation of deep brain stimulators for parkinson's disease and tremor. *Mov. Disord.* **2006**, *21*, S259–S283. [CrossRef] [PubMed]

67. Josephson, M.E. *Clinical Cardiac Electrophysiology: Techniques and Interpretations*; Lippincott Williams & Wilkins: Philadelphia, PA, USA, 2008.

68. Sigg, D.C.; Iaizzo, P.A.; Xiao, Y.-F.; He, B. *Cardiac Electrophysiology Methods and Models*; Springer Science & Business Media: Berlin, Germany, 2010.

69. Viventi, J.; Kim, D.-H.; Moss, J.D.; Kim, Y.-S.; Blanco, J.A.; Annetta, N.; Hicks, A.; Xiao, J.; Huang, Y.; Callans, D.J. A conformal, bio-interfaced class of silicon electronics for mapping cardiac electrophysiology. *Sci. Transl. Med.* **2010**, *2*, 24ra22. [CrossRef] [PubMed]

70. Merrill, D.R.; Bikson, M.; Jefferys, J.G. Electrical stimulation of excitable tissue: Design of efficacious and safe protocols. *J. Neurosci. Methods* **2005**, *141*, 171–198. [CrossRef] [PubMed]

71. Peckham, P.H.; Knutson, J.S. Functional electrical stimulation for neuromuscular applications. *Annu. Rev. Biomed. Eng.* **2005**, *7*, 327–360. [CrossRef] [PubMed]

72. Tandon, N.; Cannizzaro, C.; Chao, P.-H.G.; Maidhof, R.; Marsano, A.; Au, H.T.H.; Radisic, M.; Vunjak-Novakovic, G. Electrical stimulation systems for cardiac tissue engineering. *Nat. Protoc.* **2009**, *4*, 155. [CrossRef] [PubMed]

73. Aravanis, A.M.; Wang, L.-P.; Zhang, F.; Meltzer, L.A.; Mogri, M.Z.; Schneider, M.B.; Deisseroth, K. An optical neural interface: In vivo control of rodent motor cortex with integrated fiberoptic and optogenetic technology. *J. Neural Eng.* **2007**, *4*, S143. [CrossRef] [PubMed]

74. Duke, A.R.; Lu, H.; Jenkins, M.W.; Chiel, H.J.; Jansen, E.D. Spatial and temporal variability in response to hybrid electro-optical stimulation. *J. Neural Eng.* **2012**, *9*, 036003. [CrossRef] [PubMed]

75. Richter, C.P.; Matic, A.I.; Wells, J.D.; Jansen, E.D.; Walsh, J.T. Neural stimulation with optical radiation. *Laser Photonics Rev.* **2011**, *5*, 68–80. [CrossRef] [PubMed]

76. Wells, J.; Kao, C.; Konrad, P.; Milner, T.; Kim, J.; Mahadevan-Jansen, A.; Jansen, E.D. Biophysical mechanisms of transient optical stimulation of peripheral nerve. *Biophys. J.* **2007**, *93*, 2567–2580. [CrossRef] [PubMed]

77. Park, S.I.; Brenner, D.S.; Shin, G.; Morgan, C.D.; Copits, B.A.; Chung, H.U.; Pullen, M.Y.; Noh, K.N.; Davidson, S.; Oh, S.J. Soft, stretchable, fully implantable miniaturized optoelectronic systems for wireless optogenetics. *Nat. Biotechnol.* **2015**, *33*, 1280–1286. [CrossRef] [PubMed]

78. Robinson, R. Optogenetics sheds light on brain circuits. Is therapy next? *Neurol. Today* **2010**, *10*, 16–17. [CrossRef]

79. Williams, J.C.; Denison, T. From optogenetic technologies to neuromodulation therapies. *Sci. Transl. Med.* **2013**, *5*, 177ps6. [CrossRef] [PubMed]

80. Feiner, R.; Engel, L.; Fleischer, S.; Malki, M.; Gal, I.; Shapira, A.; Shacham-Diamand, Y.; Dvir, T. Engineered hybrid cardiac patches with multifunctional electronics for online monitoring and regulation of tissue function. *Nat. Mater.* **2016**, *15*, 679–685. [CrossRef] [PubMed]

81. Yu, K.J.; Kuzum, D.; Hwang, S.-W.; Kim, B.H.; Juul, H.; Kim, N.H.; Won, S.M.; Chiang, K.; Trumpis, M.; Richardson, A.G. Bioresorbable silicon electronics for transient spatiotemporal mapping of electrical activity from the cerebral cortex. *Nat. Mater.* **2016**, *15*, 782–791. [CrossRef] [PubMed]

82. Sun, Y.; Rogers, J.A. *Semiconductor Nanomaterials for Flexible Technologies*; Elsevier/William Andrew: Amsterdam, The Netherlands; London, UK, 2010.

83. Lee, T.; Lee, W.; Kim, S.W.; Kim, J.J.; Kim, B.S. Flexible textile strain wireless sensor functionalized with hybrid carbon nanomaterials supported ZnO nanowires with controlled aspect ratio. *Adv. Funct. Mater.* **2016**, *26*, 6206–6214. [CrossRef]

84. Gao, Q.; Dubrovskii, V.G.; Caroff, P.; Wong-Leung, J.; Li, L.; Guo, Y.; Fu, L.; Tan, H.H.; Jagadish, C. Simultaneous selective-area and vapor–liquid–solid growth of INP nanowire arrays. *Nano Lett.* **2016**, *16*, 4361–4367. [CrossRef] [PubMed]

85. Lee, W.C.; Kim, K.; Park, J.; Koo, J.; Jeong, H.Y.; Lee, H.; Weitz, D.A.; Zettl, A.; Takeuchi, S. Graphene-templated directional growth of an inorganic nanowire. *Nat. Nanotechnol.* **2015**, *10*, 423–428. [CrossRef] [PubMed]

86. Holmes, J.D.; Johnston, K.P.; Doty, R.C.; Korgel, B.A. Control of thickness and orientation of solution-grown silicon nanowires. *Science* **2000**, *287*, 1471–1473. [CrossRef] [PubMed]

87. Persson, A.I.; Larsson, M.W.; Stenström, S.; Ohlsson, B.J.; Samuelson, L.; Wallenberg, L.R. Solid-phase diffusion mechanism for GaAs nanowire growth. *Nat. Mater.* **2004**, *3*, 677–681. [CrossRef] [PubMed]

88. Zhu, B.; Niu, Z.; Wang, H.; Leow, W.R.; Wang, H.; Li, Y.; Zheng, L.; Wei, J.; Huo, F.; Chen, X. Microstructured graphene arrays for highly sensitive flexible tactile sensors. *Small* **2014**, *10*, 3625–3631. [CrossRef] [PubMed]

89. Li, X.; Yang, T.; Yang, Y.; Zhu, J.; Li, L.; Alam, F.E.; Li, X.; Wang, K.; Cheng, H.; Lin, C.T.; et al. Large-area ultrathin graphene films by single-step marangoni self-assembly for highly sensitive strain sensing application. *Adv. Funct. Mater.* **2016**, *26*, 1322–1329. [CrossRef]

90. Gong, S.; Zhao, Y.; Yap, L.W.; Shi, Q.; Wang, Y.; Bay, J.A.P.B.; Lai, D.T.H.; Uddin, H.; Cheng, W. Fabrication of highly transparent and flexible nanomesh electrode via self-assembly of ultrathin gold nanowires. *Adv. Electron. Mater.* **2016**, *2*, 1600121. [CrossRef]

91. Hong, S.Y.; Lee, Y.H.; Park, H.; Jin, S.W.; Jeong, Y.R.; Yun, J.; You, I.; Zi, G.; Ha, J.S. Stretchable active matrix temperature sensor array of polyaniline nanofibers for electronic skin. *Adv. Mater.* **2016**, *28*, 930–935. [CrossRef] [PubMed]

92. Rhyee, J.S.; Kwon, J.; Dak, P.; Kim, J.H.; Kim, S.M.; Park, J.; Hong, Y.K.; Song, W.G.; Omkaram, I.; Alam, M.A.; et al. High-mobility transistors based on large-area and highly crystalline CVD-grown MoSe$_2$ films on insulating substrates. *Adv. Mater.* **2016**, *28*, 2316–2321. [CrossRef] [PubMed]

93. Dai, X.; Messanvi, A.; Zhang, H.; Durand, C.; Eymery, J.; Bougerol, C.; Julien, F.H.; Tchernycheva, M. Flexible light-emitting diodes based on vertical nitride nanowires. *Nano Lett.* **2015**, *15*, 6958–6964. [CrossRef] [PubMed]

94. Zhang, R.; Ding, J.; Liu, C.; Yang, E.-H. Highly stretchable supercapacitors enabled by interwoven CNTs partially embedded in PDMS. *ACS Appl. Energy Mater.* **2018**. [CrossRef]

95. Ding, J.; Fu, S.; Zhang, R.; Boon, E.; Lee, W.; Fisher, F.T.; Yang, E.H. Graphene—Vertically aligned carbon nanotube hybrid on PDMS as stretchable electrodes. *Nanotechnology* **2017**, *28*, 465302.

96. Ding, J.; Fisher, F.T.; Yang, E.-H. Direct transfer of corrugated graphene sheets as stretchable electrodes. *J. Vac. Sci. Technol. B* **2016**, *34*, 051205. [CrossRef]

97. Wang, S.; Xu, J.; Wang, W.; Wang, G.-J.N.; Rastak, R.; Molina-Lopez, F.; Chung, J.W.; Niu, S.; Feig, V.R.; Lopez, J.; et al. Skin electronics from scalable fabrication of an intrinsically stretchable transistor array. *Nature* **2018**, *555*, 83–88. [CrossRef] [PubMed]

98. Lee, Y.H.; Zhang, X.Q.; Zhang, W.; Chang, M.T.; Lin, C.T.; Chang, K.D.; Yu, Y.C.; Wang, J.T.W.; Chang, C.S.; Li, L.J.; et al. Synthesis of large-area MoS$_2$ atomic layers with chemical vapor deposition. *Adv. Mater.* **2012**, *24*, 2320–2325. [CrossRef] [PubMed]

99. Zhan, Y.; Liu, Z.; Najmaei, S.; Ajayan, P.M.; Lou, J. Large-area vapor-phase growth and characterization of MoS$_2$ atomic layers on a SiO$_2$ substrate. *Small* **2012**, *8*, 966–971. [CrossRef] [PubMed]

100. Van der Zande, A.M.; Huang, P.Y.; Chenet, D.A.; Berkelbach, T.C.; You, Y.; Lee, G.-H.; Heinz, T.F.; Reichman, D.R.; Muller, D.A.; Hone, J.C. Grains and grain boundaries in highly crystalline monolayer molybdenum disulphide. *Nat. Mater.* **2013**, *12*, 554–561. [CrossRef] [PubMed]

101. Liu, K.-K.; Zhang, W.; Lee, Y.-H.; Lin, Y.-C.; Chang, M.-T.; Su, C.-Y.; Chang, C.-S.; Li, H.; Shi, Y.; Zhang, H.; et al. Growth of large-area and highly crystalline MoS$_2$ thin layers on insulating substrates. *Nano Lett.* **2012**, *12*, 1538–1544. [CrossRef] [PubMed]

102. Kim, T.-H.; Cho, K.-S.; Lee, E.K.; Lee, S.J.; Chae, J.; Kim, J.W.; Kim, D.H.; Kwon, J.-Y.; Amaratunga, G.; Lee, S.Y.; et al. Full-colour quantum dot displays fabricated by transfer printing. *Nat. Photonics* **2011**, *5*, 176–182. [CrossRef]

103. Dong-Un, J.; Jae-Sup, L.; Tae-Woong, K.; Sung-Guk, A.; Denis, S.; Young-Shin, P.; Hyung-Sik, K.; Dong-Bum, L.; Yeon-Gon, M.; Hye-Dong, K.; et al. 65.2: Distinguished paper: World-largest (6.5") flexible full color top emission AMOLED display on plastic film and its bending properties. *SID Symp. Dig. Tech. Pap.* **2009**, *40*, 983–985.

104. Park, J.-S.; Kim, T.-W.; Stryakhilev, D.; Lee, J.-S.; An, S.-G.; Pyo, Y.-S.; Lee, D.-B.; Mo, Y.G.; Jin, D.-U.; Chung, H.K. Flexible full color organic light-emitting diode display on polyimide plastic substrate driven by amorphous indium gallium zinc oxide thin-film transistors. *Appl. Phys. Lett.* **2009**, *95*, 013503. [CrossRef]

105. Choi, M.K.; Yang, J.; Hyeon, T.; Kim, D.-H. Flexible quantum dot light-emitting diodes for next-generation displays. *npj Flex. Electron.* **2018**, *2*, 10. [CrossRef]

106. Yao, S.; Zhu, Y. Nanomaterial-enabled stretchable conductors: Strategies, materials and devices. *Adv. Mater.* **2015**, *27*, 1480–1511. [CrossRef] [PubMed]

107. Sirringhaus, H.; Tessler, N.; Friend, R.H. Integrated optoelectronic devices based on conjugated polymers. *Science* **1998**, *280*, 1741–1744. [CrossRef] [PubMed]

108. Tan, Z.; Xu, J.; Zhang, C.; Zhu, T.; Zhang, F.; Hedrick, B.; Pickering, S.; Wu, J.; Su, H.; Gao, S.; et al. Colloidal nanocrystal-based light-emitting diodes fabricated on plastic toward flexible quantum dot optoelectronics. *J. Appl. Phys.* **2009**, *105*, 034312. [CrossRef]

109. Shahi, S. Flexible optoelectronics. *Nat. Photonics* **2010**, *4*, 506. [CrossRef]

110. Someya, T. Tiny lamps to illuminate the body. *Nat. Mater.* **2010**, *9*, 879. [CrossRef] [PubMed]

111. Espinosa, H.D.; Bernal, R.A.; Minary-Jolandan, M. A review of mechanical and electromechanical properties of piezoelectric nanowires. *Adv. Mater.* **2012**, *24*, 4656–4675. [CrossRef] [PubMed]

112. Wu, H.; Kong, D.; Ruan, Z.; Hsu, P.-C.; Wang, S.; Yu, Z.; Carney, T.J.; Hu, L.; Fan, S.; Cui, Y. A transparent electrode based on a metal nanotrough network. *Nat. Nanotechnol.* **2013**, *8*, 421–425. [CrossRef] [PubMed]

113. Wang, Y.; Wang, F.; He, J. Controlled fabrication and photocatalytic properties of a three-dimensional ZnO nanowire/reduced graphene oxide/CdS heterostructure on carbon cloth. *Nanoscale* **2013**, *5*, 11291–11297. [CrossRef] [PubMed]

114. Xu, J.; Wang, K.; Zu, S.-Z.; Han, B.-H.; Wei, Z. Hierarchical nanocomposites of polyaniline nanowire arrays on graphene oxide sheets with synergistic effect for energy storage. *ACS Nano* **2010**, *4*, 5019–5026. [CrossRef] [PubMed]

115. Novoselov, K.S.; Geim, A.K.; Morozov, S.V.; Jiang, D.; Zhang, Y.; Dubonos, S.V.; Grigorieva, I.V.; Firsov, A.A. Electric field effect in atomically thin carbon films. *Science* **2004**, *306*, 666–669. [CrossRef] [PubMed]

116. Geim, A.K.; Novoselov, K.S. The rise of graphene. *Nat. Mater.* **2007**, *6*, 183–191. [CrossRef] [PubMed]

117. Lee, C.; Wei, X.; Kysar, J.W.; Hone, J. Measurement of the elastic properties and intrinsic strength of monolayer graphene. *Science* **2008**, *321*, 385–388. [CrossRef] [PubMed]

118. Nair, R.R.; Blake, P.; Grigorenko, A.N.; Novoselov, K.S.; Booth, T.J.; Stauber, T.; Peres, N.M.R.; Geim, A.K. Fine structure constant defines visual transparency of graphene. *Science* **2008**, *320*, 1308. [CrossRef] [PubMed]

119. Meng, Y.; Gu, D.; Zhang, F.; Shi, Y.; Cheng, L.; Feng, D.; Wu, Z.; Chen, Z.; Wan, Y.; Stein, A.; et al. A family of highly ordered mesoporous polymer resin and carbon structures from organic–organic self-assembly. *Chem. Mater.* **2006**, *18*, 4447–4464. [CrossRef]

120. Yu, D.; Dai, L. Self-assembled graphene/carbon nanotube hybrid films for supercapacitors. *J. Phys. Chem. Lett.* **2010**, *1*, 467–470. [CrossRef]

121. Engel, M.; Small, J.P.; Steiner, M.; Freitag, M.; Green, A.A.; Hersam, M.C.; Avouris, P. Thin film nanotube transistors based on self-assembled, aligned, semiconducting carbon nanotube arrays. *ACS Nano* **2008**, *2*, 2445–2452. [CrossRef] [PubMed]

122. Margenau, H. Van der waals forces. *Rev. Mod. Phys.* **1939**, *11*, 1–35. [CrossRef]

123. Zaremba, E.; Kohn, W. Van der Waals interaction between an atom and a solid surface. *Phys. Rev. B* **1976**, *13*, 2270–2285. [CrossRef]

124. Morokuma, K. Molecular orbital studies of hydrogen bonds. III. C=O$\cdots$H–O hydrogen bond in $H_2CO\cdots H_2O$ and $H_2CO\cdots 2H_2O$. *J. Chem. Phys.* **1971**, *55*, 1236–1244. [CrossRef]

125. Janiak, C. A critical account on π–π stacking in metal complexes with aromatic nitrogen-containing ligands. *J. Chem. Soc. Dalton Trans.* **2000**, 3885–3896. [CrossRef]

126. Grimme, S. Do special noncovalent π–π stacking interactions really exist? *Angew. Chem. Int. Ed.* **2008**, *47*, 3430–3434. [CrossRef] [PubMed]

127. Maillard, M.; Motte, L.; Ngo, A.T.; Pileni, M.P. Rings and hexagons made of nanocrystals: A marangoni effect. *J. Phys. Chem. B* **2000**, *104*, 11871–11877. [CrossRef]

128. Sternling, C.V.; Scriven, L.E. Interfacial turbulence: Hydrodynamic instability and the marangoni effect. *AIChE J.* **1959**, *5*, 514–523. [CrossRef]

129. Fanton, X.; Cazabat, A.M. Spreading and instabilities induced by a solutal marangoni effect. *Langmuir* **1998**, *14*, 2554–2561. [CrossRef]

130. Hu, N.; Karube, Y.; Yan, C.; Masuda, Z.; Fukunaga, H. Tunneling effect in a polymer/carbon nanotube nanocomposite strain sensor. *Acta Mater.* **2008**, *56*, 2929–2936. [CrossRef]

131. Kovtyukhova, N.I.; Ollivier, P.J.; Martin, B.R.; Mallouk, T.E.; Chizhik, S.A.; Buzaneva, E.V.; Gorchinskiy, A.D. Layer-by-layer assembly of ultrathin composite films from micron-sized graphite oxide sheets and polycations. *Chem. Mater.* **1999**, *11*, 771–778. [CrossRef]

132. Zhang, M.; Gong, K.; Zhang, H.; Mao, L. Layer-by-layer assembled carbon nanotubes for selective determination of dopamine in the presence of ascorbic acid. *Biosens. Bioelectron.* **2005**, *20*, 1270–1276. [CrossRef] [PubMed]

133. Yang, M.; Yang, Y.; Yang, H.; Shen, G.; Yu, R. Layer-by-layer self-assembled multilayer films of carbon nanotubes and platinum nanoparticles with polyelectrolyte for the fabrication of biosensors. *Biomaterials* **2006**, *27*, 246–255. [CrossRef] [PubMed]

134. Lee, S.W.; Kim, B.-S.; Chen, S.; Shao-Horn, Y.; Hammond, P.T. Layer-by-layer assembly of all carbon nanotube ultrathin films for electrochemical applications. *J. Am. Chem. Soc.* **2009**, *131*, 671–679. [CrossRef] [PubMed]

135. Eda, G.; Fanchini, G.; Chhowalla, M. Large-area ultrathin films of reduced graphene oxide as a transparent and flexible electronic material. *Nat. Nanotechnol.* **2008**, *3*, 270–274. [CrossRef] [PubMed]
136. Compton, O.C.; Nguyen, S.T. Graphene oxide, highly reduced graphene oxide, and graphene: Versatile building blocks for carbon-based materials. *Small* **2010**, *6*, 711–723. [CrossRef] [PubMed]

micromachines

MDPI

Perspective
Liquid-Metal Enabled Droplet Circuits

Yi Ren [1] and Jing Liu [1,2,3,]* ⓘ

[1] Department of Biomedical Engineering, Tsinghua University, Beijing 100084, China;
reny14@mails.tsinghua.edu.cn
[2] Technical Institute of Physics and Chemistry, Chinese Academy of Sciences, Beijing 100190, China
[3] School of Future Technology, University of Chinese Academy of Sciences, Beijing 100049, China
* Correspondence: jliubme@tsinghua.edu.cn; Tel.: +86-10-6279-4896

Received: 13 April 2018; Accepted: 2 May 2018; Published: 5 May 2018

Abstract: Conventional electrical circuits are generally rigid in their components and working styles, which are not flexible and stretchable. As an alternative, liquid-metal-based soft electronics offer important opportunities for innovation in modern bioelectronics and electrical engineering. However, their operation in wet environments such as aqueous solution, biological tissue or allied subjects still encounters many technical challenges. Here, we propose a new conceptual electrical circuit, termed as droplet circuit, to fulfill the special needs described above. Such unconventional circuits are immersed in a solution and composed of liquid metal droplets, conductive ions or wires, such as carbon nanotubes. With specifically-designed topological or directional structures/patterns, the liquid-metal droplets composing the circuit can be discrete and disconnected from each other, while achieving the function of electron transport through conductive routes or the quantum tunneling effect. The conductive wires serve as electron transfer stations when the distance between two separate liquid-metal droplets is far beyond that which quantum tunneling effects can support. The unique advantage of the current droplet circuit lies in the fact that it allows parallel electron transport, high flexibility, self-healing, regulation and multi-point connectivity without needing to worry about the circuit break. This would extend the category of classical electrical circuits into newly emerging areas like realizing room temperature quantum computing, making brain-like intelligence or nerve–machine interface electronics, etc. The mechanisms and potential scientific issues of the droplet circuits are interpreted and future prospects in this direction are outlined.

Keywords: droplet circuits; liquid metal; quantum tunneling effect; solution electronics; electron transport; ionic conduction; quantum computing; brain-like intelligence

---

1. Introduction

Since the origin of electricity, it has become a necessity in daily life. Generally speaking, electrical circuits are rigid and continuous in their structures and components. Print circuit boards (PCB) have been commonly used in various situations (Figure 1A). However, classic rigid circuits cannot easily adapt to the human body due to poor flexibility and biocompatibility, limiting their value in the biomedical and health care fields. The increasing advancement of wearable devices and implantable systems has led to significant growth in flexible electronics (Figure 1B) [1]. Polymer nanomaterials, silk fibroin and liquid metal are being gradually adopted in soft electronics. Another main development trend in artificial circuits is molecular electronics. This was first proposed in 1974 by Aviram and Ratner [2], and refers to a field that seeks to fabricate electrical devices and circuits with single molecules and molecular monolayers [3–6]. The fabrication of molecular electrical devices includes single-molecule break junctions and molecular monolayer devices [3]. Tailored by chemical design and synthesis, the function of molecular components can be rather diverse. Deoxyribonucleic acid (DNA) has potential for molecular devices for its unique structure [5]. Figure 1C presents the logic

'AND' and 'OR' gate fabricated by DNA. Until now, molecular components including diodes, switches, memory and transistors have been intensively researched [6]. Those components can be combined to construct molecular-scale electronic computers [7]. Based on the electrical properties of the molecular diode switches, quantum mechanical calculations can also possibly be performed.

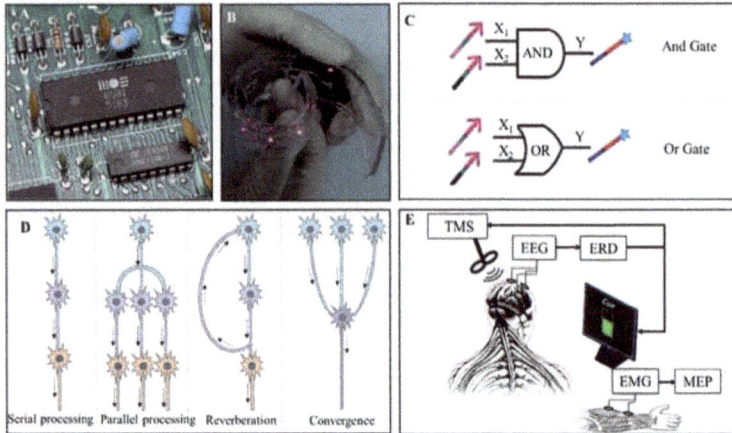

Figure 1. Respective kinds of electrical circuits. (**A**) Traditional rigid print circuit boards (PCB) circuits (https://upload.wikimedia.org/wikipedia/commons/0/0f/MOS6581_chtaube061229.jpg); (**B**) Soft electronics. Reproduced with permission from [8]; (**C**) Logic AND Gate and OR Gate based on DNA. Reproduced with permission from [6]; (**D**) Several typical types of neural circuits; (**E**) Brain–computer interface. Reproduced with permission from [9]. Note: TMS: transcranial magnetic stimulation; ERD: event-related desynchronization; MEP: motor-evoked potential; EEG: electroencephalogram; EMG: electromyogram.

Besides these artificial circuits, electrical circuits in fact intrinsically exist throughout the human body, as shown in Figure 1D. For instance, Dejean et al. recently studied the neural circuits and cell types that mediate conditioned fear expression and recovery [8]. Moreover, voltage-gated channels are important switches for signal transportation in the central and peripheral neural systems [10]. Keeping the above facts in mind and without losing generality, we can divide electrical circuits into two main categories: biologically-inspired natural circuits and human-made artificial circuits. One more trend, as indicated in Figure 1E, is that a new category is emerging to combine naturally occurring neural circuits and artificial circuits to carry out complex functions. The core of such circuits can generally be called brain–computer interfaces (BCIs) [11–14]. Clearly, BCIs are significant for patients with serious disabilities such as tetraplegia and stroke. They especially mean a lot for future human needs in extending the limits of biological capability. At this stage, BCIs based on rapid serial visual presentation have already been used to detect and recognize objects, providing a viable approach to prompt human–machine systems [14].

Despite the widespread application of currently available circuits, they all unavoidably encounter the possibility of circuit break, which severely affects the normal operation of devices. As a remedy, we here propose an unconventional concept for electrical circuits to tackle the above challenges, which we call droplet circuits. Such an electrical circuit is enabled from liquid-metal droplets (LMDs) and is conductive in discontinuous form. The structure and working style of this kind of circuit are highly analogous and similar to that of the neuro-network system. Therefore it is expected to be very useful in innovating newly emerging areas such as room temperature quantum computing, brain-like intelligence, and brain–machine interfaces, etc.

Liquid metal refers to alloys or metals with a low melting point, which can maintain their liquid phase around room temperature [15]. The electrical properties and biocompatibility of liquid metal have been proven in Yi et al.'s work [15]. In contrast to mercury, which is highly toxic, liquid metal based on gallium is relatively safe for medical applications. Liquid metal has recently been introduced into soft electronics and the biomedical field [15]. Liquid-metal sensors [16,17], memristors [18], diodes [19] and electrodes [20–22] have already been proposed for health monitoring and disease treatment (Figure 2A–C). Recently, some researches has been devoted to the study of liquid-metal droplets (Figure 2D), which show great potential in self-powered devices [23,24] and phagocytosis [25]. Yang et al. introduced millimeter-scale LMDs as thermal switches, unlocking new possible solutions for thermal management [26]. Tang et al. electrically controlled the size and rate of LMD formation [27]. Others fabricated non-stick LMDs by coating polytetrafluoroethylene particles [28] or graphene [29] on NaOH-treated LMDs. Chen et al. found that graphene-coated LMDs can be used as droplet-based floating electrodes [29]. Sivan et al. coated LMDs with n-type and p-type semiconducting nanopowders to study their electronic properties and electrochemical properties [30].

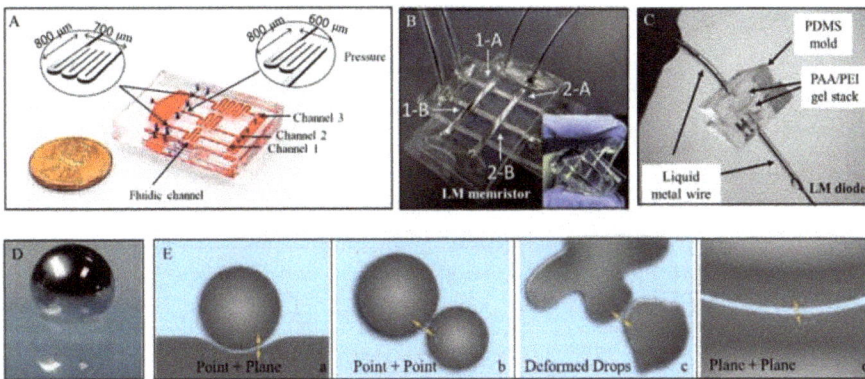

Figure 2. (**A**) Liquid-metal pressure sensor [16]; (**B**) A 2 × 2 crossbar array liquid metal memristor: 1-A, 1-B, 2-A and 2-B refer to the four bars respectively. Reproduced with permission from [18]; (**C**) Liquid-metal diode. Reproduced with permission from [19]; (**D**) Liquid-metal droplet. Reproduced with permission from [28]; (**E**) Four configurations of all-soft liquid metal quantum devices based on tunneling effect. Reproduced with permission from [31].

In contrast to previous studies, where all LMDs were contacted to realize electrical conduction, droplet circuits mean that the circuit is composed of discrete LMDs and can operate well, in the same way as traditional circuits. For this purpose, we would speculate that when the gap between the two LMDs is small enough, the electrical signal can be transported even though they are not connected. This assumption is dependent on the quantum tunneling effect theory, which is a physical phenomenon in which a micro-particle such as an electron can tunnel through a barrier that it classically could not surmount. Thus, droplet circuits can transfer electrical signals even though LMDs are separated in space. Zhao et al. have proposed a transformable soft quantum device based on liquid metal [31]. They found that liquid metal droplets can be adopted to create tunnel junctions and defined four configurations of all-soft quantum devices, as shown in Figure 2E. Based on their research, this article further explores the potential of liquid-metal droplets in constructing droplet circuits and preliminarily demonstrate its probability. Compared to traditional rigid or soft circuits, liquid-metal droplet circuits provide more flexibility without the problem of circuit breaks.

2. Electrical Conduction via the Quantum Tunneling Effect

The quantum tunneling phenomenon cannot be explained by classical mechanics, as it occurs only at the quantum scale. Compared to classical mechanics, matter in quantum mechanics has the properties of waves and particles, involving the Heisenberg uncertainty principle [32–34]. The Heisenberg uncertainty principle states that one can never exactly know the position and speed of a particle at the same time. Thus, events which seem to be impossible in classical mechanics become possible in quantum mechanics. This can be used to explain the quantum tunneling phenomenon.

Quantum tunneling is essential on many occasions, including nuclear fusion in the sun [35], astrochemistry in interstellar clouds, quantum biology, tunnel diodes (Figure 3A) [36], tunnel junctions (Figure 3B) [37], scanning tunneling microscope and quantum computing, etc. A tunnel junction is where two conductors are separated by a thin insulator to create a simple barrier between them. It can be applied to measure voltages and magnetic fields.

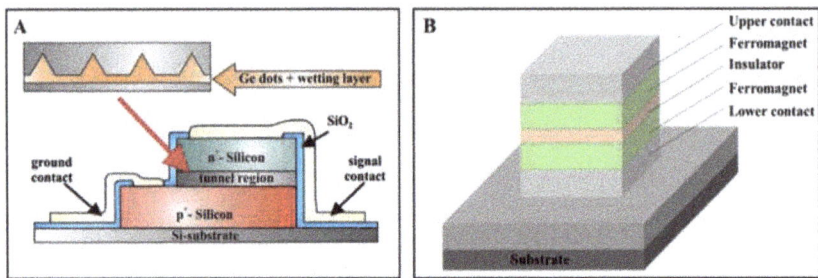

Figure 3. Illustration of the quantum tunneling effect and conventional typical application devices. (**A**) A Ge quantum dots interband tunneling diode. Reproduced with permission from [36]; (**B**) Diagram of a ferromagnet tunnel junction.

3. Fabrication of LMDs and Composition of Droplet Circuits

Until now, the methods for the fabrication of LMDs have been relatively diverse and easily available, including methods like sonication, molding, and flow-focusing [27]. Recently, Yu et al. created a method named suspension 3D printing [38], which can help quickly realize three various dimensional droplet patterns. They successfully patterned LMDs into a self-healing hydrogel (Figure 4A) and studied the relationship between the process parameters, supporting gel concentration, and the deposited micro-droplet geometry. Before that, the present lab also successfully prepared LMDs on a large scale through a fabrication method that does not use channels [39], as shown in Figure 4B. The microscopic image (Figure 4C) presents a single layer of liquid-metal micro-droplets, which were sitting together with a thin interface between each other. From the insert of Figure 4C, we can see that although the distance between the LMDs is extremely small, they are in fact discrete. Furthermore, Tian et al. proposed a microfluidic chip for liquid-metal droplet generation and sorting. Their system could manipulate these neutral liquid-metal droplets in nonconductive fluid [40]. All those studies provide viable methods for the fabrication of LMDs and the construction of droplet circuits, and more efforts should be made to improve the technology.

A rising trend in liquid metal is to cover LMDs with other materials to fabricate multifunctional circuits. For instance, liquid metal can be sealed into polymers to maintain its pattern. Water films and nanoparticles are applied to prevent the formation of an oxide layer, hold the stability and keep them discrete from each other [28,41]. Moreover, by coating LMDs with magnetic particles, they can be induced to move with the external magnetic field. Besides magnetic control, an electrical field can also induce the movement of liquid metal. Therefore, by coating specific materials, we can fabricate liquid-metal droplets with the required properties.

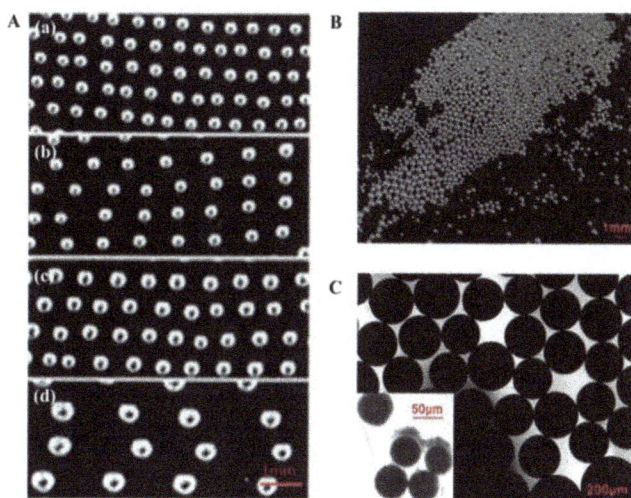

Figure 4. (**A**) Liquid-metal droplets (LMDs) floating in hydrogel with varied diameters; (**a**) 60 μm, (**b**) 90 μm, (**c**) 160 μm, (**d**) 210 μm. Reproduced with permission from [38]; (**B**) LMDs assembly fabricated through injection way. Reproduced with permission from [39]; (**C**) Microscopic image showing a single layer of liquid metal micro-droplets closely sitting together with thin interface between each other. Reproduced with permission from [39].

4. Droplet Circuits in Solution

4.1. Mechanism of Liquid-Metal-Based Droplet Circuits

Droplet circuits are mainly composed of discontinuous LMDs, ions and conductive wires such as carbon nanotubes and operate in electrolyte solution. LMDs and carbon nanotubes are mixed up and cooperate to connect the circuit electrically. Carbon nanotubes have unique electrical, thermal and chemical properties, showing great potential in nanoelectronics [42], especially for transistor applications, owing to their benign carrier mobility and velocity. Therefore, carbon nanotubes can be selected to connect droplet circuits.

Figure 5A shows the electrical conduction of droplet circuits. LMDs in droplet circuits are surrounded by carbon nanotubes and ions and electrons transfer along the pattern of LMDs with the assistance of carbon nanotubes and ions. There are two ways for LMDs to communicate in electrolyte solution, as shown in Figure 5B. When two LMDs are sitting closely enough, a quantum tunneling effect will happen and electrons can possibly transfer from one to another. If the distance of two LMDs becomes a little farther than the tunneling effect can support, carbon nanotubes and ions floating between them serve as transfer stations for electrons to flow. Through the quantum tunneling effect, carbon-nanotube transfer stations and ionic routes, the discrete LMDs are electrically connected and the whole circuit can work well. In general, LMDs are randomly arranged in electrolyte solution. If voltage is applied to the LMDs, they will be organized in order and connect the circuit (Figure 5C).

As the resistance of LMDs is smaller than other regions in electrolyte solution, the current will regularly transfer mainly along LMDs. In addition, one can change the pattern of LMDs to control the conducting direction of the current. That is, given a specific topological or directional design, the liquid-metal droplets composing the circuit can achieve the desired or regulative functions of electron transport through conductive routes or the quantum tunneling effect. Some newly emerging needs, such as room temperature computing or brain-like chips, can possibly be enabled based on such unconventional electrical circuits.

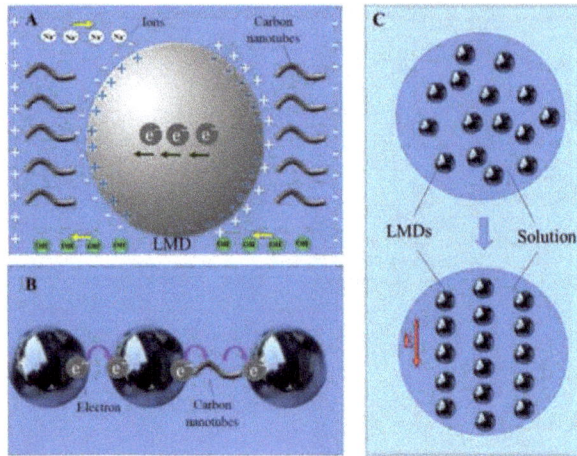

Figure 5. (**A**) The electrical environment of a LMD in NaOH solution: A LMD is surrounded by ions and conductive carbon nanotubes; tunneling effect between two LMDs; (**B**) Combination of two methods of communicating for LMDs in a circuit; (**C**) LMDs arranged with and without voltage.

4.2. Configuration of the Liquid-Metal Droplet Circuit

4.2.1. Low-Dimensional Droplet Circuits

Zero-dimensional (0D) droplet circuits refer to the case where only one LMD works. Figure 6A shows a schematic of a 0D circuit. The LMD is immersed in NaOH solution and replaces part of the wire. When the switch is closed, the lamp can be lit.

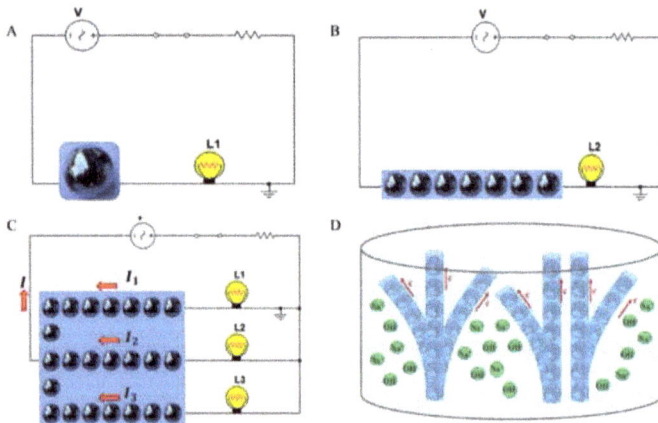

Figure 6. (**A**) Zero-dimensional (0D) droplet circuit; (**B**) 1 D droplet circuit; (**C**) 2 D droplet circuit; (**D**) Topological structure of droplet circuit.

By increasing the number of LMDs and arranging them in a line, a one-dimensional (1D) droplet circuit can be formed. The line of LMDs can be inserted into a traditional circuit as an alternative to the original wire, as shown in Figure 6B. Similarly, one can arrange LMDs in a plane with a specific pattern to realize a two-dimensional (2D) droplet circuit (Figure 6C). In principle, this circuit should also satisfy Kirchhoff's current law:

$$\sum I_{in} = \sum I_{out} \tag{1}$$

Here, I_{in} is the current flowing into a node and I_{out} is the current flowing out from the node.

Kirchhoff's law states that the charge input at a node is equal to the charge output. Therefore, the currents of the circuit in Figure 6C should meet the following equation:

$$I = I_1 + I_2 + I_3 \tag{2}$$

Due to existence of the electrolyte in the solution, charge loss is inevitable. The main loss pathways include the redox chemistry between NaOH and LMDs and the contact between NaOH and carbon nanotubes. However, these are small enough to be neglected since the conductivity of NaOH is much lower than LMDs and carbon nanotubes.

Furthermore, LMDs can be arranged in a special topological structure to fabricate three-dimensional (3D) or even dynamically transformable droplet circuits, as shown in Figure 6D. To form specific structures of 3D droplet circuits, polymer can be applied. LMDs can be sealed in polymer tubes and be arranged as required. Electrical or magnetic fields can be utilized at the same time. These 3D droplet circuits are flexible and able to transport electrons in the desired direction. Topological droplet circuits perfectly imitate the connection between neurons, which is beneficial to artificial neural connections.

The combination of liquid-metal and artificial circuits has been demonstrated before by Zhang et al. [43]. They injected liquid metal in the left and right side of the sciatic nerve near the femur of the interceptive lower part of a bullfrog body. It was found that the two electrodes successfully conducted the electrical stimuli signals to the nerves, which proves the feasibility of applying liquid metal to electrical circuits.

4.2.2. High-Dimensional Droplet Circuits

The further development of droplet circuits may be in more than three dimensions. In this regard, LMDs can be encapsulated into elastic tubes together with carbon nanotubes and electrolyte solution together to replace conventional rigid metal wires, as shown in Figure 7D. Moreover, since the pattern of LMDs can be changed by external magnetic or electrical fields, the circuit will become more fantastic and dynamically change over the time. Applying electrical or magnetic fields to LMD circuits, the structure of the circuits can be changed as required. For instance, LMDs can be induced to move by magnetic field when coating them with ferromagnetic materials (Figure 7A,B) [44]. With the help of aluminum, one can realize reliable motion control of the liquid metal droplets in the electrical field [45]. Figure 7C presents the sequential movement of a liquid-metal droplet propelled by an external electrical field. The velocity of Al/EGaIn and Ni/Al/EGaIn droplets in NaOH solution under different voltages was measured and calculated, as shown in Figure 7E [46]. Al provides a fuel source for LMDs to move, and they observed that Ni is not favorable for the electrical control of LMDs. It is apparent that the LMDs speed up with the increase of the voltage. Moreover, the current of the droplet circuits composed of LMDs in NaOH solution was relatively stable under constant voltages (Figure 7F), demonstrating promising applications of droplet circuits [47]. Recently, Isabela et al. fabricated a magnetocaloric ferrofluid based on Ga liquid metal [48]. They suspend gadolinium nanoparticles in a liquid gallium alloy and found that the material exhibited spontaneous magnetization and a large magnetocaloric effect. This is a significant progress in the magnetic control of liquid metal.

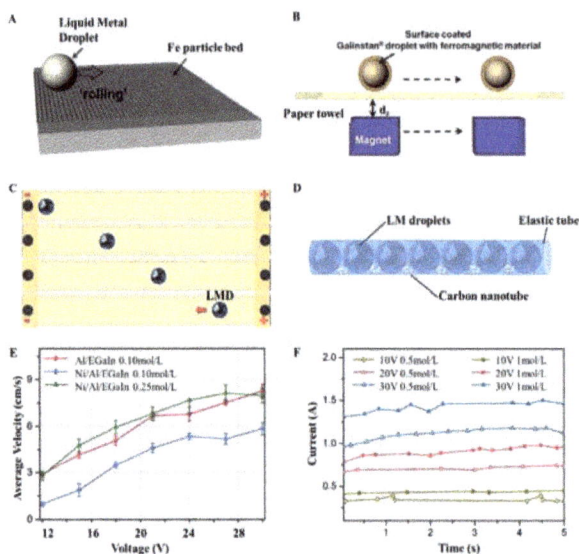

Figure 7. (**A**) Fabrication of Fe-coated LMDs. Reproduced with permission from [44]; (**B**) Surface-coated LMD with ferromagnetic materials induced by magnetic field. Reproduced with permission from [44]; (**C**) Sequential photos of LMD motion in an electrical field; (**D**) Liquid-metal droplet wire; (**E**) Average velocity of LMDs in a straight channel containing NaOH solution under exposure to different voltages, and the concentrations of the NaOH solution [46] (Reproduced with permission); (**F**) The electric current in response to different voltages, and the concentrations of the NaOH solution. Reproduced with permission from [47].

5. Discussion

This article exploits the basic features of droplet circuits and evaluates the probability of functional circuits enabled by liquid-metal droplets. Obviously, droplet circuits display unique advantages that may not be easily offered by conventional strategies. First of all, droplet circuits show a highly parallel electrical transporting capability. In addition, such circuits have excellent flexibility, self-healing capabilities and are stretchable. In contrast to existing liquid-metal soft electronics, LMD circuits consist of spatially transformable discrete liquid-metal droplets. It can tolerate greater elastic deformation and can be shaped into various electrical patterns.

Further, droplet circuits are fault-tolerant and self-error-correcting. Since LMDs can transport electrons while they are disconnected from each other, droplet circuits may not face circuit break errors. Even though the distance between LMDs can be changed by external factors such as strain and twist, carbon nanotubes will keep the circuit electrically connected. It is apparent that droplet circuits appear more robotic than traditional circuits, which is very much like the working style of a biological neurocircuit. The electrical resistance of liquid-metal droplet circuits may be higher due to the discontinuity and cross-talk between individual droplet chains, but since the quantum tunneling effect occurs within an extremely small space, the running time of the whole circuit will be greatly reduced. Particularly, droplet circuits can run in the wet environment of electrolyte solution. Electrical conduction involves not only LMDs and carbon nanotubes, but also ionic conduction. Ions in electrolyte solution play the role of electron transportation along with carbon nanotubes when the quantum tunneling effect of LMDs does not work.

Last but not least, the fabrication of liquid-metal-based droplet circuits is overall not complicated, since LMDs self-assemble in the environment of magnetic or electrical fields. The circuit configurations are in dynamic change at the same time.

From a practical point of view, due to its unique advantages, such as high flexibility and benign electrical conductivity, liquid-metal-based droplet circuits can possibly be applied to make artificial retina or cochlea as an alternative to conventional rigid wires and electrodes. Figure 8A is the presentation of a cochlear implant. The wires and electrodes should be soft and biocompatible enough to ensure minimal damage to biological tissues. Liquid-metal-enabled droplet circuits could perfectly meet these requirements. Nerve repair and neural connection is another promising development for such droplet circuits. Efforts have been made to explore the possibility of liquid-metal neural restoration. Zhang et al. employed liquid metal to reconnect the transected sciatic nerve of a bullfrog (Figure 8B) and found that the measured electroneurographic signals under electric stimulation were similar to those from the intact sciatic nerves [49]. They further put forward three types of nerve conduits to restore damaged peripheral nerves, as shown in Figure 8C. Nerve conduits combined with liquid metal can take the form of microchannels, thin slices or concentric tubes. Unlike the former trial, the current droplet circuit offers more electron transport channels, such as electrically conductive liquid metal, ionic conduction and nanowires, etc. Therefore, it would better serve medical needs. Another potential that is worth of mentioning of droplet circuits lies in its role in constructing computing chips or devices that are different from the classical framework. With discrete LMDs, droplet circuits offer opportunities to perform as quantum processors or to carry out quantum calculating, which may help the design of future quantum computers.

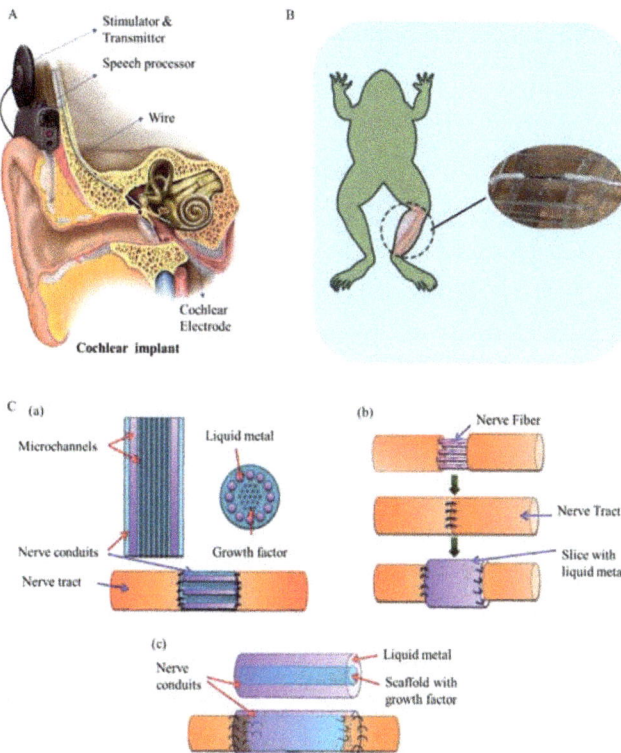

Figure 8. (**A**) The structure of a cochlear implant (https://upload.wikimedia.org/wikipedia/commons/-f/f1/Electric-Acoustic-Stimulation-EAS.png); (**B**) The transected sciatic nerve of a bullfrog reconnected by liquid metal [49]; (**C**) Three kinds of nerve conduits to repair the injured peripheral nerve: Nerve conduit with microchannels (**a**), a thin slice (**b**) and concentric tubes (**c**) [49]. (Note: Figures reproduced with permission).

6. Conclusions

This work presents a new conceptual droplet circuit that is enabled by liquid metal and demonstrates its potential role in the electronics field. The advantage of droplet circuits lies in that they address well the problem of circuit breaks and allow more flexibility and self-healing features, and show great potential for the development of smart soft electronics, which could imitate well the biological neurocircuits in nature. In addition, droplet circuits show potential for possible applications in future quantum calculation. Such droplet circuits are easily fabricated and self-error-correcting. This shows their promising value for the molding of large-scale application technologies. Further research could be conducted to fabricate stable and nano-scale liquid metal droplets in electrolyte solution and test the electrical properties of those droplets. The main difficulty of liquid-metal droplet circuits is the size of LMDs, which limits the progress of the liquid metal. Micro-scale and nano-scale circuits are currently the main trend. Compared to complementary metal oxide semiconductor (COMS) fabrication technology, the fabrication of liquid metal requires more improvement. Clearly, science, technology and applications in this direction require systematic investigation and integration in the future.

Author Contributions: Y.R. and J.L. analyzed the data and wrote the paper. J.L. conceived and supervised the project.

Acknowledgments: This work is partially supported by the National Natural Science Foundation of China (NSFC) Key Project under Grant No. 91748206, Dean's Research Funding of the Chinese Academy of Sciences and the Frontier Project of the Chinese Academy of Sciences.

Conflicts of Interest: The authors declare no conflict of interest.

References

1. Wang, X.; Liu, J. Recent advancements in liquid metal flexible printed electronics: Properties, technologies, and applications. *Micromachines* **2016**, *7*, 206. [CrossRef]
2. Aviram, A.; Ratner, M.A. Molecular rectifiers. *Chem. Phys. Lett.* **1974**, *29*, 277–283. [CrossRef]
3. Kim, Y.; Song, H. Fabrication and characterization of molecular electronic devices. *J. Nanosci. Nanotechnol.* **2015**, *15*, 921–938. [CrossRef] [PubMed]
4. Fox, M.A. Fundamentals in the design of molecular electronic devices Long-range charge carrier transport and electronic coupling. *Cheminform* **1999**, *30*, 201–207.
5. Delrosso, N.V.; Sarah, H.; Lee, S.; Nathan, D.D. A molecular circuit regenerator to implement iterative strand displacement operations. *Angew. Chem.* **2017**, *129*, 4514–4517. [CrossRef]
6. Son, J.Y.; Song, H. Molecular scale electronic devices using single molecules and molecular monolayers. *Curr. Appl. Phys.* **2013**, *13*, 1157–1171. [CrossRef]
7. Ellenbogen, J.C.; Love, J.C. Architectures for molecular electronic computers, i. logic structures and an adder designed from molecular electronic diodes. *Proc. IEEE* **2000**, *88*, 386–426. [CrossRef]
8. Wang, Q.; Yu, Y.; Yang, J.; Liu, J. Flexible electronics: Fast fabrication of flexible functional circuits based on liquid metal dual-trans printing. *Adv. Mater.* **2015**, *27*, 7109–7116. [CrossRef] [PubMed]
9. Daly, I.; Blanchard, C.; Holmes, N.P. Cortical excitability correlates with the event-related desynchronization during brain-computer interface control. *J. Neural Eng.* **2018**, *15*, 026022. [CrossRef] [PubMed]
10. Dejean, C.; Courtin, J.; Rozeske, R.R.; Bonnet, M.C.; Dousset, V.; Michelet, T.; Herry, C. Neuronal circuits for fear expression and recovery: Recent advances and potential therapeutic strategies. *Biol. Psychiatry* **2015**, *78*, 298–306. [CrossRef] [PubMed]
11. Purves, D.; Augustine, G.J.; Fitzpatrick, D.; Hall, W.C.; Lamantia, A.; Mcnamara, J.O.; Williams, S.M. *Neuroscience*, 3rd ed.; Sinauer Associates, Inc.: Sunderland, MA, USA, 2004.
12. Zhang, Y.; Wang, Y.; Zhou, G.; Jin, J.; Wang, B.; Wang, X.; Cichocki, A. Multi-kernel extreme learning machine for EEG classification in brain-computer interfaces. *Expert Syst. Appl.* **2018**, *96*, 302–310. [CrossRef]
13. Brandman, D.M.; Hosman, T.; Saab, J.; Burkhart, M.C.; Shanahan, B.E.; Ciancibello, J.G.; Sarma, A.A.; Milstein, D.J.; Vargas-Irwin, C.E.; Franco, B.; et al. Rapid calibration of an intracortical brain-computer interface for people with tetraplegia. *J. Neural Eng.* **2018**, *15*, 026007. [CrossRef] [PubMed]

14. Lees, S.; Dayan, N.; Cecotti, H.; Mccullagh, P.; Maguire, L.; Lotte, F.; Coyle, D. A review of rapid serial visual presentation-based brain-computer interfaces. *J. Neural Eng.* **2017**, *15*, 021001. [CrossRef] [PubMed]
15. Yi, L.; Liu, J. Liquid metal biomaterials: A newly emerging area to tackle modern biomedical challenges. *Int. Mater. Rev.* **2017**, *62*, 1–26. [CrossRef]
16. Jung, T.; Yang, S. Highly stable liquid metal-based pressure sensor integrated with a microfluidic channel. *Sensors* **2015**, *15*, 11823–11835. [CrossRef] [PubMed]
17. Yi, L.; Li, J.; Guo, C.; Li, L.; Liu, J. Liquid metal ink enabled rapid prototyping of electrochemical sensor for wireless glucose detection on the platform of mobile phone. *ASME J. Med. Devices* **2015**, *9*, 44507. [CrossRef]
18. Koo, H.J.; So, J.H.; Dickey, M.D.; Velev, O.D. Towards all-soft matter circuits: Prototypes of quasi-liquid devices with memristor characteristics. *Adv. Mater.* **2011**, *23*, 3559–3564. [CrossRef] [PubMed]
19. So, J.H.; Koo, H.J.; Dickey, M.D.; Velev, O.D. Ionic current rectification in soft-matter diodes with liquid-metal electrodes. *Adv. Funct. Mater.* **2012**, *22*, 625–631. [CrossRef]
20. Sun, X.; Yuan, B.; Rao, W.; Liu, J. Amorphous liquid metal electrodes enabled conformable electrochemical therapy of tumors. *Biomaterials* **2017**, *146*, 156–167. [CrossRef] [PubMed]
21. Li, J.; Guo, C.; Wang, Z.; Gao, K.; Shi, X.; Liu, J. Electrical stimulation towards melanoma therapy via liquid metal printed electronics on skin. *Clin. Transl. Med.* **2016**, *5*, 1–7. [CrossRef] [PubMed]
22. Guo, R.; Liu, J. Implantable liquid metal-based flexible neural microelectrode array and its application in recovering animal locomotion functions. *J. Micromech. Microeng.* **2017**, *27*, 104002. [CrossRef]
23. Yuan, B.; Wang, L.; Yang, X.; Ding, Y.; Tan, S.; Yi, L.; He, Z.; Liu, J. Liquid metal machine triggered violin-like wire oscillator. *Adv. Sci.* **2016**, *3*, 1600212. [CrossRef] [PubMed]
24. Zhao, X.; Tang, J.; Liu, J. Surfing liquid metal droplet on the same metal bath via electrolyte interface. *Appl. Phys. Lett.* **2017**, *111*, 101603. [CrossRef]
25. Tang, J.; Zhao, X.; Li, J.; Zhou, Y.; Liu, J. Liquid metal phagocytosis: Intermetallic wetting induced particle internalization. *Adv. Sci.* **2017**, *4*, 1700024. [CrossRef] [PubMed]
26. Yang, T.; Kwon, B.; Weisensee, P.B.; Kang, J.G.; Li, X.; Braun, P.; Miljkovic, N.; King, W.P. Millimeter-scale liquid metal droplet thermal switch. *Appl. Phys. Lett.* **2018**, *112*, 063505. [CrossRef]
27. Tang, S.Y.; Joshipura, I.D.; Lin, Y.; Kalantarzadeh, K.; Mitchell, A.; Khoshmanesh, K.; Dickey, M.D. Liquid-metal microdroplets formed dynamically with electrical control of size and rate. *Adv. Mater.* **2015**, *28*, 604–609. [CrossRef] [PubMed]
28. Chen, Y.; Liu, Z.; Zhu, D.; Handschuh-Wang, S.; Liang, S.; Yang, J.; Kong, T.; Zhou, X.; Liu, Y.; Zhou, X.C. Liquid metal droplets with high elasticity, mobility and mechanical robustness. *Mater. Horiz.* **2017**, *4*, 591–597. [CrossRef]
29. Chen, Y.; Zhou, T.; Li, Y.; Zhu, L.; Handschuh-Wang, S.; Zhu, D.; Zhou, X.; Liu, Z.; Gan, T.; Zhou, X. Robust fabrication of nonstick, noncorrosive, conductive graphene-coated liquid metal droplets for droplet-based, floating electrodes. *Adv. Funct. Mater.* **2018**, *28*, 1706277. [CrossRef]
30. Sivan, V.; Tang, S.Y.; O'Mullane, A.P.; Petersen, P.; Eshtiaghi, N.; Kalantar-Zadeh, K.; Mitchell, A. Liquid metal marbles. *Adv. Funct. Mater.* **2013**, *23*, 144–152. [CrossRef]
31. Zhao, X.; Tang, J.; Yu, Y.; Liu, J. Transformable soft quantum device based on liquid metals with sandwiched liquid junctions. *arXiv* **2017**.
32. Sen, D. The uncertainty relations in quantum mechanics. *Curr. Sci.* **2014**, *107*, 203–218.
33. Menzel, D.H.; Layzer, D. The physical principles of the quantum theory. *Philos. Sci.* **1930**, *16*, 303–324. [CrossRef]
34. Eisberg, R.; Resnick, R.; Brown, J. *Quantum Physics of Atoms, Molecules, Solids, Nuclei, and Particles*; Wiley: Hoboken, NJ, USA, 1974; p. 864.
35. Serway, R.; Serway, R. *College Physics 2*; Cengage Learning: Boston, MA, USA, 2005.
36. Oehme, M.; Karmous, A.; Sarlija, M.; Werner, J.; Kasper, E.; Schulze, J. Ge quantum dot tunneling diode with room temperature negative differential resistance. *Appl. Phys. Lett.* **2010**, *97*, 291. [CrossRef]
37. Ma, Z.; Zhou, P.; Zhang, T.; Liang, K.; Chu, P.K. Resonance-enhanced electroresistance-magnetoresistance effects in multiferroic tunnel junctions. *Mater. Res. Express* **2015**, *2*, 046303. [CrossRef]
38. Yu, Y.; Liu, F.; Zhang, R.; Liu, J. Suspension 3D printing of liquid metal into self-healing hydrogel. *Adv. Mater. Technol.* **2017**, *2*, 1700173. [CrossRef]
39. Yu, Y.; Wang, Q.; Yi, L.; Liu, J. Channelless fabrication for large-scale preparation of room temperature liquid metal droplets. *Adv. Eng. Mater.* **2013**, *16*, 255–262. [CrossRef]

40. Tian, L.; Gao, M.; Gui, L. A microfluidic chip for liquid metal droplet generation and sorting. *Micromachines* **2017**, *8*, 39. [CrossRef]

41. Ding, Y.; Liu, J. Water film coated composite liquid metal marble and its fluidic impact dynamics phenomenon. *Front. Energy* **2016**, *10*, 29–36. [CrossRef]

42. Che, Y.; Chen, H.; Gui, H.; Liu, J.; Liu, B.; Zhou, C. Review of carbon nanotube nanoelectronics and macroelectronics. *Semicond. Sci. Technol.* **2014**, *29*, 073001. [CrossRef]

43. Jin, C.; Zhang, J.; Li, X.; Yang, X.; Li, J.; Liu, J. Injectable 3-D fabrication of medical electronics at the target biological tissues. *Sci. Rep.* **2013**, *3*, 3442. [CrossRef] [PubMed]

44. Kim, D.; Lee, J.B. Magnetic-field-induced liquid metal droplet manipulation. *J. Korean Phys. Soc.* **2015**, *66*, 282–286. [CrossRef]

45. Tan, S.C.; Yuan, B.; Liu, J. Electrical method to control the running direction and speed of self-powered tiny liquid metal motors. *Proc. R. Soc. Lond. A Math. Phys. Eng. Sci.* **2015**, *471*, 32–38. [CrossRef]

46. Zhang, J.; Guo, R.; Liu, J. Self-propelled liquid metal motors steered by a magnetic or electrical field for drug delivery. *J. Mater. Chem. B* **2016**, *4*, 5349–5357. [CrossRef]

47. Zhang, J.; Sheng, L.; Liu, J. Synthetically chemical-electrical mechanism for controlling large scale reversible deformation of liquid metal objects. *Sci. Rep.* **2014**, *4*, 7116. [CrossRef] [PubMed]

48. De Castro, I.A.; Chrimes, A.F.; Zavabeti, A.; Berean, K.J.; Carey, B.J.; Zhuang, J.; Du, Y.; Dou, S.X.; Suzuki, K.; Shanks, R.A.; et al. A gallium-based magnetocaloric liquid metal ferrofluid. *Nano Lett.* **2017**, *17*, 7831–7838. [CrossRef] [PubMed]

49. Zhang, J.; Sheng, L.; Jin, C.; Liu, J. Liquid metal as connecting or functional recovery channel for the transected sciatic nerve. *arXiv* **2014**.

MDPI

St. Alban-Anlage 66

4052 Basel

Switzerland

Tel. +41 61 683 77 34

Fax +41 61 302 89 18

www.mdpi.com

Micromachines Editorial Office

E-mail: micromachines@mdpi.com

www.mdpi.com/journal/micromachines

www.ingramcontent.com/pod-product-compliance
Lightning Source LLC
Chambersburg PA
CBHW051904210326
41597CB00033B/6021